W9-CJO-202

Following
the Wild Bees

The Craft and Science of Bee Hunting

THOMAS D. SEELEY

Design and photographic assistance by Megan E. Denver
Line drawings by Margaret C. Nelson

PRINCETON UNIVERSITY PRESS
PRINCETON AND OXFORD

Published by Princeton University Press, 41 William Street,
Princeton, New Jersey 08540
In the United Kingdom: Princeton University Press, 6 Oxford Street,
Woodstock, Oxfordshire OX20 1TW

press.princeton.edu

Jacket photo by Jorik Phillips

Library of Congress Cataloging-in-Publication Data

Names: Seeley, Thomas D., author.

Title: Following the wild bees : the craft and science of bee
hunting / Thomas D. Seeley ; design and photographic assistance
by Megan E. Denver ; line drawings by Margaret C. Nelson.

Description: Princeton : Princeton University Press, 2016. | Includes
bibliographical references and index.

Identifiers: LCCN 2015038694 | ISBN 9780691170268
(hardcover : alk. paper)

Subjects: LCSH: Bee hunting.

Classification: LCC SF537 .S44 2016 | DDC 595.79/9—dc23

LC record available at http://lccn.loc.gov/2015038694

British Library Cataloging-in-Publication Data is available

This book has been composed in Sabon

Printed on acid-free paper. ∞

Printed in the United States of America

1 3 5 7 9 10 8 6 4 2

Dedicated to the honey bees

living in the Arnot Forest

 Contents

 Preface

The popularity of honey bees has skyrocketed over the past decade, so now is a perfect time to provide a book on a second way—besides beekeeping—for people to have fun with these wonderful little creatures. Our subject is an open-air sport called bee hunting. Whereas the *beekeeper* manages colonies of honey bees that are living in hives he has provided, the *bee hunter* searches for colonies of honey bees that are living in tree cavities and other homesites they have selected. The bee hunter starts his search for a wild colony by catching bees on flowers and then baiting them to forage at a small bonanza of sugar syrup that he has bewitchingly scented with anise. Next he determines the direction to the bees' secret residence from the paths of their homeward flights. Then he gradually moves his sugar-syrup feeder, together with the bees, down their flight line home—their beeline. Finally, he zeroes in on their mysterious dwelling place: some hollow tree, old building, or abandoned hive.

This all sounds rather tricky, so you might be wondering, is bee hunting something that I can do? The answer is yes. Success in bee hunting, as in all the other truly fascinating games in the world, does not require complex equipment, but it does require some special skills. This book is a guide to acquiring the simple tools and learning the ingenious methods that compose the craft of the bee hunter. While bee hunting is not a trivially easy sport, it is one that any person who enjoys spending time in nature, has patience and determination, and thrills to a treasure hunt can master and enjoy.

Bee hunting, also known as bee lining, used to be practiced widely in Europe, North America, the Middle East, and Africa. Indeed, it may be a pursuit as old as humankind, for it is likely that early humans, living in hunter-gatherer groups, searched for nests of honey bees and robbed them of brood and honey for food, as do some of the hunter-gatherer peoples who have survived to the present time. Probably the earliest written description of the methods for finding the nest of a wild honey bee colony by lining bees is that of Columella, a Roman farm owner and writer on agriculture who lived in the first century A.D. In his book on the cultivation of bees, he gives delightfully detailed instructions for capturing bees at a spring, feeding them honey, and then releasing them one-by-one to trail them back to "the lurking place of the swarm."

Within Europe, bee hunting was especially common in heavily forested regions, such as western Russia and Hungary, where it was a critical part of the craft of hollow-tree beekeeping. Forest beekeepers used various kinds of bee traps—for example, a cow-horn with a movable door in a slot and a small aperture closed with a plug—to catch one or more bees, have them load up on honey smeared inside the trap, and then release them one at a time and follow them back to their nest, usually in a hollow tree. The discoverer would carve his ownership mark in the tree's bark, cut a door in the tree's trunk to access the bees' nest cavity, and periodically climb the tree and collect some honeycombs. Bee hunting was also popular in North America following the introduction of the honey bee from Europe in the early 1600s. North American bee hunters used the same methods for finding wild colonies as had been practiced in Europe for centuries, but they rarely made repeated harvests of the honey from the colonies they found. Instead, they usually felled the trees occupied by the bees ("bee trees") and stole all their honeycombs, often killing the colony in the process.

In both Europe and North America, the importance of bee hunting gradually diminished from the 1500s to the 1900s as beekeeping with hives grouped in apiaries became increasingly common. At first, these hives were simply the hollow tree sections occupied by the bees; they were moved to the beekeeper's living place, where they functioned as log hives. The swarms from these log hives might be housed in skeps (domed hives made of twisted straw) or simple box hives. In the late 1800s, beekeepers began keeping their bees in purpose-built hives with movable wooden frames that neatly hold the bees' combs and make it possible for beekeepers to closely manage their colonies.

The invention of the movable-frame hive, along with the bee smoker, honey extractor, and other tools of modern beekeeping, has made it vastly easier for humans to get honey by keeping managed colonies in hives aggregated in apiaries rather than by hunting wild colonies in trees scattered over the landscape. So, these days, the successful bee hunter doesn't need to end his hunt by "taking up" the bee tree—cutting it down, opening it up with a sledge and wedges to expose the bees' nest, and cutting out the combs filled with honey. Instead of lugging home pails filled with white-yellow honeycombs, he can bring back a sweet haul of a different sort: delightful memories of sitting in sunny fields, watching honey bees fly off from his feeding station as they begin their flights home, and following the bees down their aerial trails to discover their mysterious abode.

The bee hunter who leaves the bees unharmed will still have an intensely pleasurable outing. After all, hunting is one of the oldest human activities, and the passion to hunt wild animals must have been, until quite recently, an immensely valuable part of human nature. I certainly feel the thrill of pursuing prey when I go bee hunting. Indeed, after making numerous moves down a beeline, closing in on a colony of forest-dwelling bees, and finally spying the glitter of the bees' wings as they dive inside their tree-cavity home, I always experience soaring feelings of success . . . even triumph! Eventually, I will head back to my own home, no richer in honey than when I started the day, for I will leave behind an undamaged bee tree and an unmolested bee colony. The dizzying jubilation of the "gotcha" moment will have passed, but alongside the delicious lingering feeling of success there will hum a quieter but equally pleasurable feeling: the satisfaction of having done no harm to the bees.

The bee hunter enjoys other rewards besides the sense of climactic exultation upon discovering the secluded home of a wild colony of honey bees. One is the fun of combining woodcraft, map and compass skills, and physical exercise to locate a bee tree, a feat that few people attempt these days. Another reward is the calming, peaceful feeling that comes from watching creatures that have evolved to serve the common good and so work together in harmony. Every beekeeper enjoys this feeling upon opening a teeming hive and beholding the thousands of residents—a queen, her daughters (workers), and her sons (drones)—living together peacefully. The bee hunter also senses the harmonious integration of the bees, and so likewise feels a quiet contentment, when he observes his wild bees standing motionless, side-by-side on his feeder comb, each one unperturbedly drinking in the sugar syrup he has provided. There is no shoving or fighting for the rich forage. The bee hunter again witnesses the bees' amazing communal functioning when they recruit their nest mates to his feeder. He introduces only a handful of foragers to his sugar-syrup extravaganza, but these few often become tens or hundreds in an hour or so, as the first bees recruit a task force of colony mates to help harvest the bounty the bee hunter is providing. How are these little insects, each one possessing a brain smaller than a grass seed, able to communicate so effectively? And how do they not get lost when commuting often a mile or more across hill and dale between their home and the bee hunter's feeder?

Anyone who gives bee hunting a try will quickly observe, firsthand and close-up, that there is much about honey bees that is mysterious, especially their astonishing gifts of communication and navigation. So for some who go bee hunting, the greatest reward will be that it puts a bee in your bonnet to stop, watch, and ponder the marvelous six-legged beauties that help keep our planet flowering and fruitful. For others, the greatest pleasure from hunting for a wild colony of honey bees will be that it is a wonderful way to get away from things, both physically and mentally. When you are outdoors on a weekend morning, in a sunny meadow filled with wildflowers, with your eyes tracking bees as they fly away from your feeder and with your brain figuring which individuals will return there soon, it is easy to avoid thinking about your work and other personal threads—things good and bad, important and trivial. Bee hunting is a great way to escape your own head, because the bees, which are among the most astonishing creatures on the planet, can grab your full attention.

One thing about honey bees that seems to be in everyone's head is the fact that although these fascinating insects are small and cute, they do pack a punch. So you might be wondering about the danger of being stung while bee hunting. It is certainly true that bees will be buzzing around the bee hunter while he is tending his sugar-syrup feeder and making observations. This may be terrifying to a newcomer to the sport, but I can emphatically state that there is no danger of being stung while bee hunting; the bees are entirely friendly to the human, who is providing them a free lunch. They will fight off a yellow jacket wasp if she finds the feeder. But the bees have no reason to sting the bee hunter, and I've never been stung in my nearly 40 years of bee hunting. One can get stung only if there is a careless accident, such as putting a bare arm down on a bee resting on the armrest of your folding lawn chair, or if there is a thoughtless act, such as slapping at a bee flying near your face. It may seem incredible to the novice, but barring such an accident, there is virtually no danger of being stung while bee hunting.

There is one more general point about bee hunting that needs to be stressed at the outset: it is a sport that is equally enjoyable and suitable for both men and women. The bee hunter is referred to as "he" throughout this book, but this is done merely for ease of reading. Therefore, every "he," "him," and "his" in this book also encompasses a "she," "her," and "hers."

Besides being a guide to the sport of bee hunting, this book is an expression of my gratitude to a gentleman whom I never met, but who taught me how to find wild colonies of honey bees, a skill that has helped me greatly in my scientific studies of how honey bees live in nature. I am referring to George H. Edgell (1887–1954). He was a professor at Harvard University, a director of the Museum of Fine Arts, Boston, and the author of three landmark books on the history of architecture. These are all breathtaking accomplishments. My gratitude toward him, though, derives from something he did that was rather modest: he wrote a small book on bee hunting titled *The Bee Hunter*. Dr. Edgell published this work in 1949, late in his life, by which time he had some 50 summers of experience as a bee hunter. His death in 1954 followed my birth by only two years. So, inevitably, he and I have worked apart, but I like to think that if our lives had overlapped more, then we'd have worked together from the heart.

We have both hunted and found dozens of bee trees in the wooded hills of New England and New York. We have also both shared what we have learned about the craft of bee hunting by writing a little book on the subject. There is also the curious fact that we have both pored over the typewritten manuscript for his book, *The Bee Hunter*, he to make final corrections with his fountain pen, me to study his craft. This manuscript, by the way, is now archived in the Special Collections vault in the Mann Library of Cornell University. Finally, we have both owned one very special copy of *The Bee Hunter*. This is the copy that Dr. Edgell gave to Karl von Frisch around 1950, shortly after he had deciphered the honey bee's famous waggle dance. This is the communication behavior—a ritualized reenactment of a flight to a profitable food source—that a successful forager bee performs inside the hive to tell her nest mates where to find flowers brimming with super-sweet nectar or laden with fresh pollen. On the first page of this particular copy of his book, Dr. Edgell wrote the following inscription, "To Dr. Karl von Frisch, with the homage of an amateur to a great scientist. G. H. Edgell." In 1982, shortly before his death, Professor von Frisch passed this book on to his foremost student Professor Martin Lindauer, who in turn passed it on to me in 2002. Probably in 2022, I will pass it on to the Cornell University library to become part of the Special Collections in Mann Library. So, eventually this little globe-trotter will settle down, in utmost safety, alongside its lovely "larval stage," Edgell's marked-up manuscript.

Besides *The Bee Hunter* by Edgell, and the present work, there are various other books in which the activity of bee hunting is described. Some are works of fiction, such as *The Oak Openings*, by James Fenimore Cooper (1848), and *The Bee Hunter*, by Christopher Brant (1966). These storybooks contain highly simplified descriptions of how bee hunting is done, probably because they are second-hand or third-hand accounts of the process. As for the nonfiction books on the subject, many, including *Following the Bee Line* by Josephine Morse (1931) and *The Appalachian Chronicles* by Andrew J. Smith (2010), also suffer from being written by people who have never gone bee hunting. No one could find a bee tree by pursuing the methods they describe.

This leaves a literal handful of books on bee hunting that are written by authors who have actually hunted bees and found bee trees. These include *Birds and Bees* by John S. Burroughs

(1875), *Bee Hunting* by John R. Lockard (1908), *The Bee Hunter* by George H. Edgell (1949), and *Hunting Wild Bees* by Robert E. Donovan (1980). The present book claims membership in this special group, for it is based on nearly 40 years of bee-hunting experience in various locations in New York and New England, and in places farther afield such as the jungle-covered mountains of Thailand. Over the years, I have lined my way to more than 50 wild colonies of honey bees. Within this tiny constellation of books written by veteran bee hunters, the present work hopes to shine brightly by being not only a reliable "how-to" book but also a revealing "how-come" book. In other words, besides describing the methods of bee hunting, I will report in "biology boxes" at the ends of the chapters what biologists have learned about the remarkable behavioral skills of honey bees that the bee hunter observes when he induces them to lead him to their home. In short, I will present both the hows and the whys of the sport of bee hunting.

I have many acknowledgments to make. Thanks in greatest measure go to friend and fellow bee hunter Megan E. Denver, who gave me the idea of writing this book and urged me to pursue it. Her inspiration and encouragement, along with her help in reviewing the literature on bee hunting, shooting and editing the photographs, and planning the book's design, has been generous and constant. Megan's partner, Jorik Phillips, kindly helped with the field photography in the Arnot Forest, so I am grateful for his friendship and contributions as well. Deep thanks are also due Margaret C. Nelson, who created all the line drawings for this book. Her skill at converting my sketches on graph paper into crisp computer images underlies much of the visual appeal of this work.

I am also highly grateful to several fellow field biologists whose work and friendship has been a source of inspiration over the 40 years that I have been studying wild colonies of honey bees. Special thanks go to my earliest partner in bee hunting, Kirk Visscher, who designed a simple but effective box for bee hunting (see chapter 2) and who was my "study buddy" when I began learning the craft of bee hunting back in 1978. I extend heartfelt thanks as well to several students who have helped me survey the wild colonies of honey bees living in the Arnot Forest: Koos Biesmeijer, Barrett Klein, David Tarpy, and Michael Smith, but especially Sean Griffin, who is truly a whiz at bee hunting. I also feel indebted to several fellow Cornellians, including my close colleague in the Department of Neurobiology and Behavior, Paul Sherman, who shares my love of studying animals in nature; and my friends in the Department of Natural Resources, Peter Smallidge (director of the Arnot Forest) and Donald Schaufler (manager of the Arnot Forest), who allow me work whenever, wherever, and pretty much however I want in the Arnot Forest. I hope that this little book will serve as a compelling statement on the special value of this magnificent woodland, if ever such testimony is needed.

My appreciation extends to the Cornell University Agricultural Experiment Station, whose financial support over the years has supported my research on the honey bee colonies living in the wild in the Arnot Forest.

I am also deeply grateful to several people, including Ann Chilcott, Bernd Heinrich, Maira Seeley, Robin Seeley, and Mark Winston, for providing extremely valuable feedback on early drafts of the manuscript. Three gifted photographers of insects—Helga Heilmann, Kenneth Lorenzen, and Alexander Wild—provided the close-up photographs of the bees, which help immensely to bring the story to life. To these three individuals, and to the Pierpont Morgan Library in New York City, which provided the image of Henry David Thoreau's journal entry on bee hunting, I give special thanks for their material assistance.

To the staff at Princeton University Press, I owe a constantly increasing debt. To Alison Kalett, editor for biology and earth science, belongs credit for patience, support, and guidance on writing for a broad audience. I also owe Betsy Blumenthal, Nathan Carr, and Carmina Alvarez debts that are hard to liquidate for managing the pieces of my manuscript and then finding a way to transform them into a handsome book. Lastly, I am deeply grateful to Patricia Fogarty for her persistence and skill in copyediting the manuscript.

To all who have contributed to this project, I extend my most cordial thanks.

Tom Seeley
Ithaca, New York

Following
the Wild Bees

 CHAPTER 1

Introduction

This book is about bee hunting—a fascinating open-air sport in which you find a flower patch humming with honey bees; you capture, sumptuously feed, and release a dozen or so of these bees; and then, using simple equipment but sophisticated skills, you trail these bees, step-by-step and in whatever direction they fly, back to their home. Bee hunting is a sport of infinite variety. If you start a hunt where colonies of bees living in the hives of beekeepers are fairly common, such as a suburban neighborhood or a country district with farms, then you might find yourself zeroing in on somebody's apiary. But if you start someplace wilder, say along an uninhabited road running between wooded mountains, then you'll probably find yourself following a beeline for the deep woods, homing in on the one tree out of the thousands around that is the secret residence of a wild colony of bees (fig. 1.1). Wherever this outdoor game is played, it combines almost everything that is desirable in a sport: it requires no costly equipment, can be played alone or in a group, exercises both the muscles and the brain, demands skill and persistence, builds suspense, and ends in either harmless disappointment or exhilarating triumph.

The greatest thrill in bee hunting, for most bee hunters, is to locate a wild colony of bees living in a stately tree deep in a

Fig. 1.1. Bee tree, with a knothole that serves as the nest entrance visible in the left trunk.

forest, and to sense the colony's vitality by watching the heavy traffic of its foragers zipping in and out of a picturesque knothole. Whenever I have this experience, I am reminded of the opening words of Aldo Leopold's classic tribute to nature, *A Sand County Almanac*: "There are some who can live without wild things, and some who cannot." Like most beekeepers, I love the honey bee colonies that I keep in my hives, for they are easily observed and studied. But I am *in* love with the honey bee colonies that live in the woods. They choose by themselves their tree-cavity homes, build as they see fit their beeswax combs (fig. 1.2), gather all their nourishment from flowers in the surrounding landscape, and fight without aid every predator or disease that crosses their lives. In short, these wild colonies draw fully upon the wonderful array of structural, physiological, and behavioral adaptations that constitute the biology of honey bees.

Wherever there are honey bees, there exist both managed colonies living in beekeepers' hives and wild colonies living in tree cavities, rock clefts, and the walls of buildings. While it is true that managed and wild honey bee colonies lead rather different lives—the former are manipulated to produce honey and pollinate crops, whereas the latter are left alone and can do whatever boosts their survival and reproduction—the bees in both types of colonies are virtually identical. The members of these two groups look, function, and act so similarly because the two groups have essentially the same genetic composition. This genetic similarity is a consequence of the frequent swapping of genes between the managed and wild colonies living in the same geographical area. Part of this genetic exchange between the two groups arises because the colonies living in beekeepers' hives produce swarms that escape and then lead lives in the wild, while at the same time the colonies living in natural abodes produce swarms that beekeepers collect and then install in their hives.

FIG. 1.2. The nest inside the bee tree shown in fig. 1.1. The tree trunk housing the nest has been split open, revealing the beeswax combs containing honey (above) and brood (below). On the left side of the cavity, about two-thirds of the way up, is the entrance opening. Total height of the nest is 5 feet.

The exchange of genes between managed and wild colonies also takes place in a second, more sensational way: the curious sexual behavior of honey bees. Every queen bee mates on the wing with 15–20 males drawn from the neighboring colonies living within four or so miles from her home. This shameless promiscuity of queen honey bees evolved because high genetic diversity among a queen bee's female offspring—that is, the workers in her colony—is essential to her colony's health. These days, it also has the effect of blending the genes in the managed and the wild colonies living in the same region. Incidentally, this extensive gene flow between managed and wild colonies explains why humans haven't created distinct breeds of honey bees through selective breeding, analogous to what has been done in the domestication of dogs, horses, and sheep.

It is a remarkable fact that humans have been keeping honey bees for at least 9,000 years, starting in the Middle East, and yet this insect still remains an essentially wild animal. The honey bees residing in beekeepers' hives look and behave the same as their wild counterparts. Indeed, they are all as much at home in a hollow tree as in a manufactured hive, and they are all fully capable of surviving entirely on their own.

"TO FAIR HAVEN POND—BEE HUNTING"

This little treatise is written in conscious admiration of Henry David Thoreau—not that Thoreau did much bee hunting. He did, however, write a remarkably detailed and reliable description of how it is done. I will try to do likewise, while bringing things up to date. Thoreau's guide to bee hunting is tucked away in the two-million-word journal—the daily record of things he thought, saw, and felt—that he kept from 1838, shortly after leaving Harvard College, to 1861, one year before his death. The entry of special interest to bee hunters is the one made on September 30, 1852, which begins "10 AM to Fair Haven Pond—Bee Hunting. Pratt, Rice, Hastings & myself, in

Fig. 1.3. Start of Thoreau's journal entry for Thursday, September 30, 1852. It reads as follows: "10 AM to Fair Haven Pond—Bee Hunting. Pratt, Rice, Hastings & myself in a wagon. A fine clear day after the coolest night and severest frost we have had. The apparatus was, first a simple round tin box about 4½ inches in diameter and 1½ inches deep, containing a piece of empty honey comb of its own size and form filling it within ⅓ of an inch of the top."

a wagon." (fig. 1.3). It runs over eight pages, making it one of Thoreau's longer entries for all of that year. What makes it such a trustworthy guide to the sport of bee hunting is that Thoreau does not include anything that he has been told but has not seen. There is no hearsay. Instead, Thoreau sticks to recounting what he saw and what he did on that "fine clear day" when he and the cobbler Hastings climbed on a wagon with two experienced bee hunters, Minot Pratt and Reuben Rice, and rode out to a field beside Fair Haven Pond, some two miles south of the village center in Concord, Massachusetts.

Thoreau starts by describing a bee hunter's most important piece of gear: the bee box. This is a smallish, usually wooden, two-chambered box that is immensely useful in the critical first stage of every hunt, when the bee hunter must convince a dozen or so foraging bees to quit visiting flowers and accept instead a tantalizing free lunch. The lunch counter is usually a piece of old beeswax comb filled with either diluted honey or sugar syrup lightly scented with anise extract. The bee box used by Thoreau's company consisted of a "round tin box about 4½ inches in diameter and 1½ inches deep, containing a piece of honey comb of its own size and form" together with a wooden box that would be set atop the tin one. Thoreau tells how the bee hunters first caught several bees in the wooden box and then, after setting this box atop the tin box, gently opened an escape hatch in the wooden box's floor to allow the trapped bees to climb out and find the irresistible bait below. A few minutes later, the wooden box was lifted gently off the tin box, freeing the bees to fly home when each had taken her fill.

With a bee box in hand, one is ready to start bee hunting, and Thoreau describes how they searched for honey bees on the flowers by Fair Haven Pond, but found none there. The goldenrod flowers (*Solidago* spp.) were withered from a severe frost the previous night, and the purple aster flowers (*Aster nova-angliae*) were sparse. After eating lunch, the four men headed back to the village along Walden Road. When they reached Walden Pond they noticed fresh goldenrod and purple aster flowers on the sunny hillside sloping from the roadside down to the pond (fig. 1.4). These flowers were "resounding with the hum of bees." The team quickly captured and sent forth some dozen honey bees, each one laden with diluted honey drunk from Pratt's (or Rice's) bee box. The bees flew off in three directions, all toward places where the men knew there were colonies living in hives, not toward the forest homes of wild honey bee colonies.

Pratt was probably disappointed by where the bees were going, for he knew that a wild honey bee colony represented real treasure for its first finder. Indeed, he had told Thoreau about this earlier in the year. In the February 10, 1852, entry of his journal, where Thoreau records his discovery of a colony of bees living in a hemlock tree beside Fair Haven Pond, he also mentions that "Pratt says ... I may get five dollars for the swarm [colony], and perhaps a good deal of honey." Thoreau,

Fig. 1.4. Worker honey bee collecting nectar and pollen from purple aster (*Aster nova-angliae*) flowers.

though, shows no disappointment about their sunny September day spent in bee hunting. Indeed, in summarizing his feelings about the day, he wrote, "I feel the richer for this experience. It taught me that even the insects in my path are not loafers, but have their special errands. Not merely and vaguely in this world, but in this hour, each is about its business. If, then, there are any sweet flowers still lingering on the hillside, it is known to the bees both of the forest and the village. The botanist should make interest with the bees if he would know when the flowers open and when they close."

Thoreau's account of bee hunting in mid-19th-century New England depicts not just the sport, but also the author: a poet-naturalist who loved the uninhabited roads that led away from Concord, to the fields, woods, swamps, and ponds where he found the wild things that he enjoyed, preferably by himself. The bee hunter who, like Thoreau, delights in observing the ways of nature, perhaps especially in solitude, will love how the sport of bee hunting will lead him to places of quiet, natural beauty, ones that he would never discover were he not lining bees back to their unknown dwelling places.

Thoreau also liked to see how little money it is possible to spend, by working with one's hands and simple tools, and still complete a project. For instance, he built his cabin by Walden Pond for $28.12½, a respectably low price even in 1845, when (as Thoreau proudly records in his journal) the annual rent for a mere dormitory room at Harvard was $30.00. He achieved this economy by doing things like borrowing an axe and using it to fell young white pines and hew them into house timbers, rather than going to a sawmill and buying what he needed for his sills, corner posts, studs, and rafters. We shall see that a bee hunter who already has a watch, a magnetic compass, and some scrap lumber, and is handy with woodworking tools (or has a friend who is), can kit himself out for this sport for less than $28.12½.

BECOMING A BEE HUNTER

Like Thoreau, I learned the basics of bee hunting from an old-timer who lived in Massachusetts. His name was Dr. George Harold Edgell. He was both a distinguished professor of architectural history at Harvard University and an avid bee hunter at his summer place in New Hampshire. His obituary in the *New York Times*, on June 30, 1954, notes that over his career, he wrote four books: *A History of Architecture*, *The American Architecture of Today*, *A History of Sienese Painting*, and *The Bee Hunter* (fig. 1.5). The latter is a trim little book of 49 pages that was published by Harvard University Press in 1949.

The Bee Hunter is a gem. In it, Edgell introduces himself as a successful bee hunter of 50 years' experience. He also explains on page 1 that his main source of motivation to write this little book is the irritation he has felt in reading various books and articles on bee hunting, *all* of them written by people who must have never gone bee hunting. The telltale sign of their lack of firsthand experience is that the methods they describe could not possibly work. (It seems Edgell did not know about Thoreau's journal entry.) I think Edgell vented a bit of his

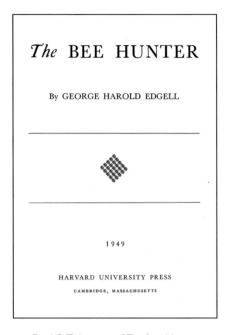

The BEE HUNTER

By GEORGE HAROLD EDGELL

1949

HARVARD UNIVERSITY PRESS
CAMBRIDGE, MASSACHUSETTS

FIG. 1.5. Title page of *The Bee Hunter*.

irritation with these fakers when he wrote, "It is time for someone who has hunted bees and found bee trees to write the facts."

Edgell further introduces himself to his readers by explaining that his interest in this sport began at the age of 10, when he was initiated "by an old Adirondacker who had sunk to driving his grandfather's mules in Newport, New Hampshire. George Smith, as I shall call him, was a character, to the youngster as fabulous as Paul Bunyan. He took his whiskey neat. He smoked and chewed at the same time and could spit without removing the pipe from his mouth. His profanity would take the bluing off a gun barrel. Withal, he was one of the kindest and most generous of men and a mighty bee hunter before the Lord, or devil if one prefers."

I discovered *The Bee Hunter* in the summer of 1978, when I returned to my family's home near Ithaca, New York, with a

PhD in biology and was on the lookout for something new to study. Ever since high school, I had been passionately interested in honey bees, and for my doctoral thesis I had enjoyed figuring out how the scout bees in honey bee swarms evaluate prospective nesting cavities, so there was no question that I'd keep going with the bees. I was feeling then, as I still do today, a strong desire to better understand how these beautiful little creatures live as wild colonies in forests, rather than as managed colonies in apiaries. Unless I could learn how *Apis mellifera* lives in its natural environment, I would never truly understand how its physiology, behavior, and social life adapt it to the natural world.

It seemed to me that one of the most profound environmental changes that beekeepers impose on their bees is the crowding of colonies in apiaries. In Europe, the original home of the honey bees we have in North America, this change started around 200 A.D., when people began to switch from *hunting* for colonies living in tree cavities to *keeping* colonies in purpose-made hives, which at first were simply hollow logs and inverted baskets. This switch made it possible to pack honey bee colonies together in apiaries, which of course makes beekeeping practical for humans. Unfortunately, living under crowded conditions can also make life hard for the bees, just as it can for us. Colonies of honey bees living jam-packed in an apiary endure greater competition for food, a higher likelihood of having their honey stolen, and an elevated risk of catching infectious diseases.

I also suspected that the difference in spacing between managed and wild colonies might be startlingly large. On the one hand, I knew that beekeepers (including me) usually space their hives just a few feet apart. On the other hand, I had just read the remarkable book by Dorothea Galton, *Survey of a Thousand Years of Beekeeping in Russia*, in which she describes how, in medieval Russia, the honey bees inhabiting trees in the forests around the city of Nizhny Novgorod had a density of only four or five colonies per square mile, which meant that the average distance between colonies was approximately half a mile—more than 2,500 feet! I wondered, are the wild colonies living in the forests in North America also spaced so widely?

Coming back to Ithaca, which is also home to Cornell University, was exciting because I knew that close by was an ideal natural area in which to find the answer to my question. Fifteen miles southwest of Ithaca is a 4,500-acre research forest, the Arnot Forest, owned by Cornell (fig. 1.6). The rugged land adjoining the Arnot Forest, which includes the Newfield and Cliffside State Forests, is also largely forested, having been protected by New York State or abandoned by agriculture during the past one hundred years. The whole area is a natural haven for the study of wildlife, including wild honey bees. I had fallen in love with the Arnot Forest a few years before when I had installed bait hives (nest boxes mounted in trees to capture honey bee swarms) of different sizes in the forest, to determine the bees' preferred volume for a nesting cavity, and to this day it is one of my favorite outdoor haunts. Now I was eager to see if I could map the nests of the wild honey bee colonies living in the Arnot Forest and so learn about their dispersion across this vast, hilly, forested landscape.

Step one was to read up on bee hunting. A quick search of the card catalog in Mann Library—the enormous library for biology, agriculture, and applied social sciences at Cornell—revealed two books under the subject heading "Bee Hunting." Great!

The first book that I tracked down in the library's stacks was a thin paperback of 72 pages, smaller than my hand, with a title that suggested it might be, despite its size, a comprehensive handbook: *Bee Hunting: A Book of Valuable Information for Bee Hunters—Tells How to Line Bees to Trees, Etc.* Published in 1908, it was written by John R. Lockard (1858–?). It seems that Lockard was a kindly gentleman who had lived somewhere in the mountains of West Virginia, Kentucky, or Tennes-

Fig. 1.6. A view of the Arnot Forest, as seen from a lookout point along Irish Hill Road. Photo taken in early October, near the peak of the autumnal colors.

see. He explains in the preface that his book is a distillation of his knowledge of bee hunting gained during "forty years in nature's school room," and was written to "inculcate a desire for manly pastime and make [the reader's] life brighter." He definitely succeeded in both aims with me, for I finished his book feeling both keener to get out hunting and more optimistic of success. I had learned several valuable bits of the bee hunter's craft, including the importance of closely examining every tree, stump, or log when you think you are near a colony's hidden home; how a bee's flight path away from your bait can easily deviate from a direct course home unless you are making sightings in a large clearing; and what a stroke of good fortune it is to discover bees collecting water along a stream or other wet spot in the woods, for this reveals that a colony lives nearby.

However, because Mr. Lockard provides only a fuzzy description of how to zero in on a bee tree once you've determined its general direction, and because his technique for introducing bees to an irresistible free lunch involved kindling a fire, heating a flat stone, and melting a piece of honeycomb to produce aromas enticing to bees, reading his book didn't leave me feeling ready to start mapping the bee trees in the Arnot Forest. Not only did I still lack a clear sense of the mechanics of lining bees, I knew that Mr. Al Fontana, the no-nonsense manager of the Arnot Forest at the time, would throw me out of the place, probably forever, if he caught me lighting fires here and there in "his" lovely forest.

The second book I discovered that morning in Mann Library was George H. Edgell's small masterpiece, *The Bee Hunter*. Almost immediately after pulling it from the shelf, I knew I had found the handbook I'd been seeking. In 45 pages of text, one line drawing (of a bee box), and seven black-and-white photos, Edgell explains how to build a bee box, what additional pieces of paraphernalia are useful when hunting bees, which seasonal conditions favor success in bee hunting, how to establish a line to the bees' nest, how to execute a series of moves along the line to zero in on the bees' home, how it can be a major challenge ultimately to find the specific tree in which the bees are living, and, if one desires, how to "take up" (that is, cut down and extract honey from) a bee tree. I read and re-read this delightful little book at least four times that day, partly the better to absorb Edgell's wonderfully detailed instructions, but also for the sheer pleasure of reading his charming writing. Consider, for example, his description of how "sparingly" one should use oil of anise when scenting a sugar syrup bait: "When I say sparingly, I mean more than the word ordinarily implies. The cork of the anise bottle rubbed on the comb and the comb then licked with the tongue will provide anise enough for one's purpose. More will make the bees quite drunk, they will refuse to suck but buzz around looking for the anise and eventually retire to the flowers to sober up." By the end of the afternoon, I knew I was ready to take action.

FIRST BEE HUNT

When I returned to Ithaca for the summer of 1978, I lived at my parents' house, but I worked at the Dyce Laboratory for Honey Bee Studies, which is part of the Department of Entomology at Cornell. With me was my close friend Kirk Visscher, who had just moved to Cornell to start his graduate studies in entomology. As was often our habit—we'd studied together at Harvard, and there I'd supervised Kirk's undergraduate thesis project—we began kicking around thoughts on important mysteries regarding honey bees that we could address through our research. Both Kirk and I had become fascinated by honey bees in high school, and we were both keen to tackle exciting questions about their behavior and social life. I shared with Kirk my question about the spacing of honey bee colonies in the wild and my plan of answering it by becoming (I hoped) a crackerjack bee hunter and locating all the colonies in the Arnot Forest. I also shared proudly my discovery of Edgell's charming little book.

By the following morning, Kirk, who is enviably intelligent and skilled at building gadgets, had read the book and built a bee box. It was inspired by the one Edgell shows in *The Bee Hunter*, but was simpler in design and thus easier to build. And it works just as well, if not perhaps better! It is such a good design that I'm still using a sturdy bee box that I built 35 years ago based on Kirk's design from that evening. We shall meet this bee box design in chapter 2.

It took Kirk and me another couple of days to assemble the rest of the equipment needed for bee hunting: an opaque cloth for covering the bee box, two squares of empty comb cut from a frame of sturdy brood comb and sized to fit loosely inside the bee box, a pint-size canning jar filled with sugar syrup "sparingly" scented with anise extract, a dropper bottle for filling the comb cells with the syrup, a small bottle of paint and a fine brush for labeling bees, a magnetic compass, a topographic

FIG. 1.7. Entrance sign for the Arnot Forest.

map of the area, a roll of plastic flagging for marking the route into (and out of) the woods, a watch, a notebook and pencils, and a toolbox or backpack for carrying everything. Also useful but not essential are a wooden crate, to serve as a stand for the bee box, and a folding chair, to make it comfortable to tend the bee box at the start of the hunt.

It is a 45-minute drive from the Dyce Lab to the entrance of the Arnot Forest (fig. 1.7), so when Kirk and I motored there in one of the lab's small fleet of green Chevy pickup trucks, we had time to discuss where we should make our first attempt at bee hunting. We decided to begin high in the forest, near its center and hence miles from any houses, to minimize the risk of engaging bees from a beekeeper's hive. It was a hot and sunny day, perfect weather for honey bees to be out foraging, so we assumed it would not be hard to find bees on flowers. It was also the middle of July, so we were making our first attempt when the summer honey flows (times of plentiful nectar) in our region were finished. Our major sources of honey in spring and early summer are the black locust, sumac, and basswood trees, together with raspberry bushes and various herbaceous plants, such as dandelions and white clover.

Kirk and I did not yet recognize the importance of bee hunting when nectar is not available in abundance from natural sources. Only when the bees cannot find plentiful nectar in flowers will they forage enthusiastically from a square of dark, old, beeswax comb filled with sugar syrup. Evidently, taking syrup from a bee hunter's comb feels to the bees like robbing honey from another colony's nest, which is dangerous work. Robbers are often caught and killed inside the hives they are plundering. Therefore, the only times when the bees are wild about a bee hunter's bait comb are days when the weather is fair but they cannot find good, safe sources of carbohydrate, namely flowers chock-full of nectar. A warm day following a night with heavy frosts, like what Thoreau described for September 30, 1852, is ideal for bee hunting, but any fair day between spring and fall when there are flowers in bloom is worth a try (see chapter 3).

We parked atop Irish Hill, beside an old field, one of many cleared from the forest by Irish immigrants who established farms on this hill in the 1800s. They grew crops and grazed their stock on these fields until the mid-1930s, when the Federal Resettlement Administration—part of President Franklin D. Roosevelt's New Deal—helped farmers living on submarginal land, such as the thin soils on Irish Hill, move to more productive farmland. Propped on my writing desk is a rusty license plate from the last days when people lived up here. I found it in the weeds while checking a bait hive that I had mounted in a white pine behind the cellar hole of one of the long-gone farmhouses. This rectangle of disintegrating sheet steel has "NY 33" (for New York State 1933) stamped in small type at the top, and "4J79-63" (the vehicle registration number) stamped in large type below. I wonder if its owner knew, when he got this license plate back in 1933, how soon he'd be leaving his hilltop farm.

Kirk and I wandered about the site for half an hour, looking for honey bees, rather surprised that we could not find any. Finally, Kirk spotted a worker bee collecting pollen on a bush of multiflora rose (*Rosa multiflora*) and captured her in the bee box by gently maneuvering the flower bearing the bee into the open end of the box and then snapping the door shut. He then lured the bee to the box's back chamber, by letting light enter the window in the rear wall of this chamber. Finally, he locked her in the back chamber by closing the sliding divider that partitions the box into two chambers. Meanwhile, using the dropper bottle, I had filled one of the small squares of empty comb with sugar syrup and passed it to Kirk to put in the box's front chamber. Once this had been done, he gently set the bee box atop the wooden crate we'd brought along as a stand, covered the window in the back chamber, raised the sliding divider between front and back chambers (so the bee could find the comb), and put the thick cloth over the box to darken it. If the bee had seen light shining through any cracks, she would have struggled to escape at these places rather than crawl all about and bump into our syrup-filled comb.

We waited five minutes, to give her plenty of time to discover our wonderful surprise, and then Kirk opened the box's door slowly and smoothly, to avoid disturbing the bee. Our bee had indeed found the syrup (hurray!) but hadn't finished loading. Peering in, we saw her standing motionless on the comb, concentrating on the task of sucking up syrup. In about a minute, once she had filled herself to capacity, she walked out of the box into the sunlight, groomed a speck of syrup off a wingtip, probably warmed her flight muscles by shivering, and finally took flight. We crouched, the better to see her against the blue sky as she slowly circled around the bee box, gradually expanding her movements into figure eights, mostly in the eastern direction from the feeder. Of course, we tried our hardest to follow her convoluted flight, hoping to see her eventually streak off in a beeline that would reveal the direction of her home, but we lost her from view while she was still circling. It looked like her home was more east than west, but that is about all we could say. The big questions now: Would she return for another load of the syrup bait? And would she share with her hive mates the news of her discovery?

Kirk and I waited hopefully, nervously, and patiently, reciting a key sentence from Edgell's book: "The most important quality for a successful bee hunter is patience." To help "our" bee find the comb again and to make it easier to label her if she should return, we moved the comb from its shady location just inside the bee box to a sunny spot just outside the box. After 9 minutes and 20 seconds, we heard the familiar sound of a honey bee! Then we saw her, presumably "our" bee, first approaching the comb but then darting off, next whizzing in circles around us, then poising in flight just above the comb, but finally landing and then standing atop the comb with her wings folded and her tongue extended, imbibing more of our irresistible bait. The bee was ours!

We applied a dot of light green paint on the abdomen of our bee, which transformed her into a little friend named Green Abdomen (recorded as "G-ab" in the notebook). Fortunately, Green Abdomen was not frightened by encountering a small square of old comb that was filled with anise-scented "honey" and was sitting oddly in bright sunshine in front of a wooden box on the side of a dirt road, for she kept coming and going from the bee box, and soon she had even brought companions (fig. 1.8). Most settled quickly on the comb. We had a six-color paint set, and it was not long before we had 12 bees labeled for individual identification, each daubed with a dot of one of the six colors on abdomen or thorax. We had also begun recording data in our notebook on two things of huge interest to a bee hunter: the bees' vanishing bearings when they flew home, and their departure and return times. The latter would enable us to estimate the distance to the bees' home.

FIG. I.8. The first bee has spread the word to her nest mates about a "free lunch" of sugar syrup.

Once the bees became accustomed to our feeder, they began to circle less and less when leaving the feeder, so it became easier for us to keep our eyes locked on a bee while she flew away. But we were just beginners, and we found it frustratingly difficult to get sightings we could trust. After about an hour of this work, we had accumulated 14 sightings of bees that flew off straight enough that we could follow them for 50 to 100 yards. The compass readings for these 14 flights home were 90°, 86°, 84°, 82°, 79°, 72°, 74°, 81°, 98°, 93°, 87°, 90°, 81°, and 92°. The average of these readings is 85°, so there could be no doubt; the home of these bees lay nearly due east.

Getting data on the busy bees' departure and return times was much easier, and we quickly learned that the length of time a bee was gone from the feeder ranged widely, from 7 minutes 50 seconds to 13 minutes 40 seconds. We noticed, however, that the four shortest "away times" were all about 8 minutes.

Knowing that these bees had flown home, unloaded their syrup, and flown back to the bee box, all in 8 minutes, meant that we could make an estimate of the distance to their home. We knew that bees fly about 15 miles per hour (hence 4 minutes per mile, or about the speed of a human sprinter), and we estimated that a bee needs about 2 minutes inside the nest to unload the droplet of syrup she has brought home; we had seen that this was how long our bees were taking to load up. Subtracting an estimated 2 minutes of unloading time from the 8 minutes of total time gone gave 6 minutes of flight time total, and therefore approximately 3 minutes of flight time for each leg of the journey. Using this estimate of 3 minutes of flight coming or going, together with the estimated flight speed of 4 minutes per mile, we calculated a flight distance to the nest of approximately three-quarters of a mile (3 minutes of flight/4 minutes per mile = ¾ mile).

Three-quarters of a mile. Yikes! Stretching out in the valley to the east, and sweeping up the hillside beyond for nearly 2 miles, was a mature hardwood forest filled with thousands of majestic sugar and red maple, beech, black and yellow birch, shagbark and pignut hickory, red and white oak, tulip poplar, and white ash trees, plus dense stands of hemlock trees on the north-facing slopes, and a sprinkling of towering white pines. These woods had been logged heavily in the late 1800s, but since then they had been left alone, and now they were packed with trees large enough to enclose a cavity of the sort a honey bee colony selects for its home. We wondered, would we ever be able to find the one tree out there that was home to Green Abdomen and her housemates?

To try to answer this question, we followed Edgell's advice about making a move down a beeline. We refilled our comb with syrup and placed it back inside the bee box. The bees were upset and acted suspiciously. One by one, though, they began landing again, so eventually it was standing room only on our matchbook-size square of comb, with each of the 15 or so bees again fixated on filling her honey stomach with our tantalizing bait. At this point, we softly closed the door of the bee box, slipped a rubber band around it to keep the door shut, packed up our gear, and moved 100 yards in the direction the bees had flown home. This took us to the brushy edge of the old field. Edgell describes making initial moves down the beeline of 300 or 400 yards, but that seemed far too daring to us. Also, we weren't ready to plunge into the forest.

When we gently opened the bee box in the new spot, some of the bees rushed out and disappeared without circling, but several were still loading, and when they departed they did so calmly and seemed to take their bearings, for they circled slowly about the new location before they too vanished. Now we were nervous. Would any of the bees return to this new site and so make the move, or would they all return to the original spot and be lost? We noted the time, and waited, knowing that probably nobody would show up for at least 10 minutes. The wait felt endless. But after 12 minutes, we heard the wonderful, silvery tone of a honey bee returning to our comb. In a couple more minutes, other bees had also landed on the comb, including Green Abdomen. Yippeee! Kirk and I felt we were really on our way to becoming, like George Smith, "a mighty bee hunter before the Lord."

Now, however, we also had to face facts: we needed to move deeper into the woods. We scouted down the beeline and found, as expected, only unbroken forest. We recalled Edgell's somber words: "Released in the woods, a bee circles up into the trees and disappears. Sometimes it is hard to tell whether [she] goes forward or back." We soon learned, however, that we could make moves to spots in the woods where there was a gap in the canopy, usually where a big tree had fallen, and in these places we could watch the bees circle up and disappear more on one side of the opening than the others. This helped us stay on the line, and gradually we drew nearer the bee tree. As we did so, more bees buzzed around us, and we had to refill the comb every few minutes. Also, the labeled bees were spending less and less time away between stops at the feeder.

Eventually, on our sixth move, made on our second day, we jumped the operation ahead by about 150 yards, to a large canopy opening where the forest floor fell away sharply and a huge red oak, which had grown on the edge of this drop-off, had fallen downhill. The bees moved to this site easily; a few even seemed to fly along with us as we made this move. Soon we had dozens of bees buzzing about us, and we were refilling the comb faster than ever, so we figured the tree must be near. But the bees were behaving oddly when they departed: instead of launching forward along the eastward line we had been following, they were circling away in all directions. Eventually, we realized that our bees, once they reached the top of the canopy opening, were flying north, not east, which told us that the bee tree was off to the left of the line we'd been traveling along, not farther ahead.

This was a valuable observation, for it meant that we did not have to work down the steep hillside that fell away before us; instead, we could work around on the shelf above it. It also meant that the bee tree was nearby, hence only half a mile from our starting point, not the three-quarters of a mile that we had originally estimated. And finally, it meant that now we needed to change our hunting tactics, switching from moving down a beeline to conducting a tree-to-tree search. This was a new challenge, one that felt almost hopeless because it seemed there must be hundreds of trees that we would need to inspect closely, bottom to top, to find the home address of our bees. We took heart, though, from another timeless tip from Edgell: "It is only a matter now of looking carefully enough to discover the tree." Undeniably true, but a bee tree can be exasperatingly difficult to find, as we shall see in a later chapter, when I describe one bee tree that took me three years to find. But on this day, we were beginners and we enjoyed some beginners' luck. We fanned out in the direction the departing bees were heading,

Fig. 1.9. The bee tree that the author and a fellow bee hunter, Kirk Visscher, found on their first bee hunt.

and after about an hour of going from tree to tree, scanning up and down each tree's trunk for the flash of flying bees, Kirk shouted, "Found it!" Indeed he had. Some 20 feet up the northwest side of an 18-inch diameter (at breast height) hemlock tree was a knothole with honey bees pouring in and out (fig. 1.9). At that moment, for these two novice bee hunters, there could have been no more splendid sight.

BIOLOGY BOX 1
How Abundant Are Wild Colonies of Honey Bees?

Honey bees have been living in the forests of eastern North America for some 400 years, following their introduction from Europe by English settlers starting in the 1620s, and perhaps also by Spanish settlers starting even earlier (Sheppard 1989). The hunting of the wild colonies of these bees, for their honey, probably began soon after they were brought to the New World. Already in 1720, a Mr. Paul Dudley published in the Philosophical Transactions of the Royal Society of London a letter titled "An account of a method lately found out in New-England for discovering where the bees hive in the woods, in order to get their honey." By the early 1800s, various writers, including Washington Irving (1835), were saying that bee hunting was a pleasurable and profitable pursuit for frontiersmen, since the honey obtained by plundering the nests of wild colonies was not just a treat but was also easily bartered and sold. In the journals of the Lewis and Clark Expedition, we find the following note by William Clark for Sunday, March 25, 1804, shortly after the expedition party had left St. Louis and was camped along the Kansas River: "river rose 14 Inch last night, the men find numbers of Bee Trees, & take great quantities of honey" (Moulton 2002). And in 1925, Thompson described "a most successful bee tree hunter," Lester Shaw, who lived in rural Potter County in northern Pennsylvania. He had collected over 1,700 pounds of honey from bee trees in a single season and could count by the hundreds the wild colonies he had found. It seems clear that in the not-so-distant past there were count-less skilled bee hunters with deep knowledge of the abundance of wild colonies of honey bees in North America. Unfortunately, what these hunters knew was not documented and therefore was lost when they died. So when Kirk Visscher and I began to investigate this subject in the summer of 1978, we felt we were looking, both figuratively and literally, out across new scientific terrain.

Kirk and I had difficulty finding honey bees on flowers in the Arnot Forest in early July, so we decided to postpone further bee hunting until late August. (This was a wise decision, for reasons to be explained in chapter 3.) We knew that the dense stands of goldenrod (*Solidago* spp.) plants sprouting up beside the roads and in the abandoned fields in the Arnot Forest would start blooming in late summer. We also knew that the flowers of goldenrod, being superabundant, are the pri-mary late-summer source of nectar and pollen for bees living in this part of New York State, so we figured that it would be easy to find worker bees buzzing around on these plants in late August. By then, though, Kirk was busy with his coursework as a new graduate student at Cornell, so I conducted the rest of the bee hunting on my own. I too would need to step back from this project, in mid-September, to return to Cambridge, Massachusetts, for the first dinner meeting of the Soci-ety of Fellows at Harvard, which was supporting my postdoctoral stud-ies. But until then I was free to go bee hunting all day and every day. Sweet!

I hunted the wild honey bees in the Arnot Forest from August 26 to September 13. Nearly every day was hot and sunny. This gave the bees plenty of time to be out working the flowers, and it gave me an excel-lent opportunity to be out hunting the bees. I would arrive in the dewy fields before the sun and would continue hunting until the bees stopped visiting my bait comb at dusk. The midpoint of each day was marked by the sound of the noon whistle rising up from the Cotton-Hanlin sawmill down in the hamlet of Cayuta, a mile and a half from the forest's western boundary.

To make the most of the 19 days I had available for the task, I focused my hunting in the southern and western parts of the forest, where, in abandoned pastures and along the railroad line that skirts the forest's south boundary, I found 17 open spots where I could see in all

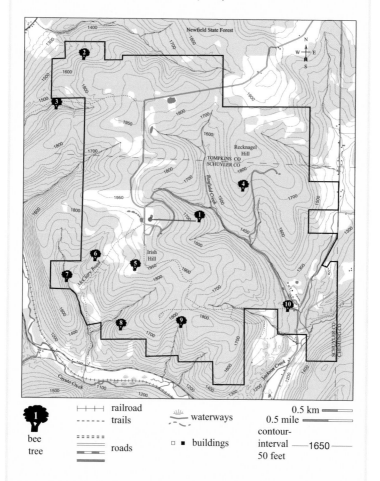

Map of the Arnot Forest showing the locations of the ten bee trees found there in 1978. The site of each bee tree is marked by the base of a bee-tree symbol. Red line denotes the path of the author's first bee hunt.

directions; I also found lush stands of goldenrod plants, whose shining flowers bobbed with honey bees. At each location, I could easily fill my bee box with bees to start a beeline, and I could get good readings of the vanishing bearings of the bees when they finished loading up on my sugar syrup and flew home to their nests. As shown in the figure, these beelines steered me to 10 bee tree colonies, 9 living in the Arnot Forest and I just outside its western boundary.

Almost certainly, the nine colonies that I found were not all of the colonies living in the Arnot Forest. After all, I did not establish beelines from flower patches located in the northern and eastern regions of the forest, which make up about 50% of its total area. I estimated, therefore, that the 9 colonies that I found inside the Arnot Forest's boundaries were only about half of the colonies living in this forest; hence there were about 18 colonies total living there in the fall of 1978. Given that the area of the Arnot Forest is almost exactly 7 square miles, this estimate of 18 colonies total meant that the abundance, or density, of the wild colonies in this forest back in 1978 was approximately 2.6 colonies per square mile; this translates to 1 colony per square kilometer.

This first estimate of the abundance of wild colonies of honey bees living in the Arnot Forest has proven to be a good general estimate for this location. Since 1978, I have made two more surveys of the colonies living in this forest. In 2002 and 2011, I located 8 and 10 colonies, respectively, living inside (or just outside) the forest's boundaries (Seeley 2003a, 2007; Seeley et al. 2015). In conducting the two additional surveys, I again covered about 50% of the Arnot Forest's total area, so my estimates of the total number of wild colonies living in the Arnot Forest in these two surveys are 16 and 20 colonies. Therefore, the estimates of the density of these wild colonies for the 2002 and 2011 surveys are 2.3 and 2.9 colonies per square mile, close to the estimate for the 1978 survey of 2.6 colonies per square mile. This consistency in the surveys' results gives me confidence that 2 or 3 colonies per square mile (or about 1 colony per square kilometer) is a reliable estimate of the abundance of wild honey bee colonies living in the Arnot Forest.

This estimate of 2 or 3 wild colonies per square mile in this rugged, heavily forested region of southern New York State appears to fall at the lower end of the range of wild-colony densities that have been reported for *Apis mellifera* across its vast range, which includes Europe, the Middle East, and Africa (where it is native) and the Americas and Australia (where it is introduced). As reviewed by Hinson et al. (2015), the published estimates of wild colony densities range from 0.3 to 20 or more colonies per square mile in both natural (for example, nature preserves) and agricultural habitats. It seems clear, therefore, that bee hunters will enjoy good hunting for wild colonies everywhere on the planet where these cosmopolitan bees live. I can add that in all 10 places outside of the Arnot Forest but still in the United States where I have gone bee hunting over the last 20 years—eastern and central New York, western Pennsylvania, northern Connecticut, western Massachusetts, central Vermont, and various locations in Maine—I have had no difficulty establishing beelines that led me to wild colonies. Moreover, I have spotted wild colonies of honey bees living in trees and buildings while exploring in the following European countries: Ireland, Great Britain, Sweden, France, Switzerland, Germany, and Austria. Given the results of the studies reviewed by Hinson et al. and my personal experiences, I am confident that if you give bee hunting a try at the right time of the year (that is, whenever honey bees are foraging) and in a place with nest cavities for honey bees (wherever there are trees and buildings), then you will have success in establishing a beeline that can lead you back to a wild colony of honey bees.

CHAPTER 2

The Bee Box and Other Tools

One of the charms of bee hunting is how few tools you need to find a bee tree. I suspect this simplicity is one reason Thoreau was so delighted by the day he spent bee hunting at Fair Haven Pond and Walden Pond. Another attraction of the sport is the compact size of the bee hunter's tools. The complete toolkit of a bee hunter fits easily into a knapsack in the field and a shoebox back at home.

The bee hunter's most important tool is the bee box. This is the ingenious little wooden box with two compartments that enables you to capture (literally) the attention of a small number of foraging bees. You do so by briefly locking up a half dozen or so foragers in a dark space with nothing to do except crawl about and stumble upon a breathtaking source of bee treasure: an unguarded square of beeswax comb brimming with sugar syrup.

You can make the bee box yourself if you are handy with woodworking tools, or you can have one built from specifications by a friend or cabinetmaker if you are not. The box should be of a size that you can hold easily in one hand. My bee box is 7½ inches long, 4 inches wide, and 3½ inches high. This is pretty big as bee boxes go, but it works for me because I have large hands. These dimensions can easily be shrunk to

FIG. 2.1. Author's bee box, with the sliding divider raised and propped open. A square of beeswax comb, with one bee atop it, sits in the front compartment.

5½ inches long, 3½ inches wide, and 3 inches high (fig. 2.1). It is fine to use scrap lumber for building your bee box, but the wood should be clean, for honey bees are repelled by the odors of oil, tobacco, and the like.

The bee box is built with two compartments that are connected by an opening that you open and close by manipulating a vertical sliding divider (fig. 2.2). The front compartment opens to the outside and is closable with a hinged door. The door hinges must be securely anchored to both the door and the floor of the box to withstand the rough handling the door will receive when you slap it shut to trap a bee. The back compartment has a rear wall that is made either of glass or clear acrylic plastic and is covered with a wooden slide that can be raised, when needed, to admit light to the back compartment. You admit this light after trapping a forager in the front compartment—by snapping the front door shut around a flower that has a foraging bee on it—and then raising the sliding di-

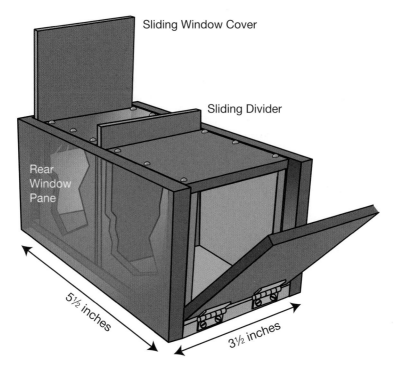

Sliding Window Cover

Sliding Divider

Rear
Window
Pane

5½ inches

3½ inches

FIG. 2.2. Cutaway diagram of a bee box, showing the window in the rear of the back compartment and the sliding window cover.

vider. The bee tries to escape toward the light coming in the window in the back compartment, and then you trap her there by lowering the sliding divider. Now you are ready to open the door of the front compartment to catch another bee. In this way, you can collect five or more bees in the back compartment of the bee box, which is the critical first step in establishing a beeline.

The bee box should be tightly and sturdily constructed and then varnished inside and out so the wood is waterproofed. It is best to do so with spar varnish, which is made to protect the spars and other wooden parts of boats that are clear-coated rather than painted. Protecting your bee box in this way is important because it is not uncommon for the box to be left outside in all sorts of weather. Sometimes, for example, you will have started a line, gotten the bees running well, and made several moves down the line, but not reached the bees' home by the end of the day. In this situation, you can leave the bees and pick them up again the next morning to continue your hunt. Before leaving the bees and your bee box for the night, you fill with sugar syrup both of your two small squares of comb, and you tuck one in the front compartment (leaving the door open) so that the syrup in this square of comb can't be diluted by rain or dew. The other comb is left outside the bee box to provide additional bait.

If nicely built, your bee box will last as long as you do. And over the years, it will work better and better, for the more it is weathered and the more it smells of bees and beeswax, the more the bees will like it.

Once you have a bee box, it is easy to put together the rest of the items in a bee hunter's toolbox. Here is what else you will need to conduct this rather cozy sport.

An opaque cloth

This is for covering the bee box to darken its interior after you have jailed a small band of bees inside it. Darkening the box will help your prisoners find the comb loaded with sugar syrup that you have placed inside the box. I'm still using the 12- by 15-inch scrap of shiny orange velvet that I found in my mother's rag box 37 years ago, when I started bee hunting. And I'm very glad I still have this gaudy cloth cover for my bee box, for it helps me make folks laugh when I demonstrate my craft. The laughter comes at the point in my presentation when I carefully cloak my bee box using this snazzy cloth. Folks know I'm a hardcore biologist, so I suppose it looks funny to see me doing something that is a standard part of a magician's act. Truthfully, though, I do feel rather like a magician at this critical first stage of a bee hunt, because I know that inside my special little box I am converting bees that have zero interest in my bee-

Fɪɢ. 2.3. A medicine dropper is useful for filling a piece of old comb with sugar syrup.

hunting endeavor into bees that are hooked on my ambrosial bait. Indeed, soon each one will be zooming to and from my inexhaustible mother lode, airlifting home loads of bee gold, and will continue doing so until either she wears out or I discover her dwelling place.

Two squares of empty comb

You will need two small squares of empty comb, cut to a size that will slide easily into the front compartment of your bee box. The ones that I use are about 2 inches by 2 inches. They will work best if they come from an old, dark, brood comb whose cells are blackened and strengthened from repeated use (fig. 2.3). These can be obtained from a beekeeper.

A jar of sugar syrup

Your bee-hunting syrup should be highly concentrated and lightly scented. I make up mine in pint-size canning jars, which

I recommend because they are wonderfully leak-proof. The preparation of the syrup is easy: put 1½ cups of pure, white cane sugar (330 grams of sucrose) in a pint jar and then pour in enough boiling water to make 2 cups (475 milliliters) of golden syrup. Getting the sugar fully dissolved requires about 5 minutes of stirring. Finally, add 1 drop of anise extract (see biology box 2) to your pint of warm syrup.

A small dropper bottle

This is very handy for neatly filling the cells in your comb squares with your sugar syrup. Mine is a 1-ounce amber bottle with a screw cap. Having the bottle in which to store the dropper when you are finished with it is extremely handy, because it means that the rest of your equipment will not get sticky.

A bottle of anise extract, a jar lid, and a small piece of 8-mesh hardware cloth

These three items will help make the site of your feeder comb highly conspicuous to the bees that have attended the dances of bees reporting your food source, have found their way to the general area of your feeder, and now are endeavoring to find the source of this amazing anise-scented "nectar." The metal jar lid that I use is from a 1-pound jar of peanut butter, and my piece of 8-mesh hardware cloth (a sturdy screen material made of galvanized wire with a 1/8-inch mesh; sold in hardware stores) is cut large enough to cover this jar lid. I have stapled the hardware cloth over a hole cut in a brightly painted square of wood that I place over the jar lid (fig. 2.4). To give the feeding station a strong scent of anise, in addition to that released from the scented-syrup held in the comb, I pour several drops of the anise extract into the jar lid (open side up) and then place the hardware-cloth over the lid so my bees won't get soiled with the extract. I then set the syrup-filled comb atop the hardware cloth over the jar lid. Now the bees that have been dispatched to my aromatic treasure trove can find it easily. A

FIG. 2.4. Setup for conspicuously marking the feeder comb's location with scent and color. The comb sits on a painted screen board covering a jar lid that holds a few drops of anise extract. Everything sits atop a small, brightly colored table.

note on timing: I set up my anise-extract-loaded and hardware-cloth-covered jar lid during the 5 or so minutes that I wait before releasing the half-dozen bees that I have captured in my bee box. This ensures that the feeder comb will be well marked with scent when these bees, and the nest-mates that they recruit, make their first attempts to find the feeding station.

A set of paint pens

These are for labeling bees for individual identification. When I began bee hunting, I used little bottles of typewriter correction fluid of various colors, but these have become hard to find. Sometimes, out of nostalgia, I will use one of the sets of shellac paints and tiny camel's hair brushes that I used for many years in studying the social organization of foraging in honey bees (described in Seeley, 1995). But now I find that paint pens and watercolor paints are the most practical way to apply identification marks to the bees that are zipping back and forth between my conspicuous feeder and their hidden home (fig. 2.5).

Having bees labeled so they can be recognized as individuals is necessary, of course, for timing a bee to see how long she is gone from the comb, which in turn is essential for estimating the distance to the bee's residence.

A watch

The watch should have a second hand, so you can time with reasonable precision how long individual bees are gone from your feeding station.

A magnetic compass

Ideally, this will be one that is designed for making accurate directional sightings, including the vanishing bearings of your homeward-flying bees. I use a Cammenga model 27 Lensatic compass, which is extremely durable and gives accurate sightings.

FIG. 2.5. *Left:* Flagging, sighting compass, and topographical map. *Right:* Paint pens and watercolor paint set, for labeling bees for individual identification.

A notebook and pens or pencils

I recommend that you carry a sturdy, bound notebook in which to record data on the bees' departure and arrival times, vanishing bearings, and other details for each of the sites where you stop as you work down a beeline. I also suggest that at each site you make a table of the trip times and the departure directions of individual bees. Doing so will allow you to quickly learn which individuals are your most dedicated and speediest foragers and hence your best sources of information on the distance to your bees' home. You will also quickly see whether the traffic of bees leaving your combs forms more than one aerial trail, and so you will quickly learn whether your bees are coming from more than one colony.

A knapsack or toolbox

This is for carrying your bee-hunting tools together with your water, food, sunscreen, and so forth (fig. 2.6). It is best to leave your insect repellent at home, however, because it can disturb the bees. A knapsack has the advantage of leaving your hands free for carrying the bee box (and perhaps a stand, see below) when you make a move down a beeline, but a toolbox makes everything readily accessible.

Fig. 2.6. Bee hunter's toolbox, holding bee box, cover cloth, jar of sugar syrup, dropper bottle, paint set, magnetic compass, topographic map, and other tools.

TABLE 2.1 Bee-Hunting Tools

Tools Needed

Bee box	Set of paint pens
Opaque cloth	Watch
Two squares of empty comb	Magnetic compass
Jar of sugar syrup	Notebook and pens or pencils
Small dropper bottle	Bottle of anise extract, a jar lid, and a
Knapsack or toolbox	small piece of 8-mesh hardware cloth

Optional Equipment

Stand for the bee box	Roll of vinyl flagging
Folding chair	Topographic map of the area

OPTIONAL EQUIPMENT

Besides the essential items listed above, you may want to use several optional items as well.

A stand for the bee box

This makes the tending of your bee box and syrup-filled combs much easier than if they are set on the ground. Thoreau does not mention whether he, Pratt, Rice, and Hastings used a stand, but they might have built a simple one using rocks pilfered from a stone wall. Edgell certainly used a stand, for he describes using one "made of an upright piece of wood such as a four-foot section of a rake handle with a flat board nailed on top and the lower end sharpened so it can easily be thrust in the ground." As for me, sometimes I've used an empty hive body (the wooden box of a bee hive) set on end. Other times, I've used a small table that I built specially for bee hunting. It is just 20 inches tall and has a small, 9- by 11-inch top, so it is easy to carry around. It is also painted yellow, so it functions as a conspicuous landmark for the bees. I find that having some sort of stand for the bee box makes it a pleasure to label bees, take notes on their departure and return times, and get sightings of their vanishing bearings, especially if I have also brought along a comfortable folding chair.

A folding chair

This is definitely a handy item, especially if you are beginning your hunt near your car or truck so it is easy to carry the chair to your starting point. I like to use a chair, but generally I do so only at the start of a hunt, when I'm likely to spend an hour or so patiently extracting information from the comings and goings of the bees. At this point one is, of course, starting from scratch in getting solid information about the direction(s) and the distance(s) the bees are traveling to reach home. The initial site is generally the one where you will need to make your most numerous sightings and take your most detailed notes. You will do these things most effectively while sitting comfortably.

A roll of vinyl flagging

If you are hunting someplace where you might get lost, then flagging tree branches every hundred feet or so is a must. Flagging has been especially valuable to me when my hunt requires an overnight break and the next morning I need to find my way back to where I left the bee box the evening before. There have been times when I might have lost my bee box if I had not marked my route through the woods by tying fluorescent pink flags to tree branches as I hiked back to my car.

A topographic map of the area

In principle, a topographic map from the U.S. Geological Survey (or other relevant organization if you are outside the United States) is optional, but in practice it is nearly essential, at least for me. Almost always, I will pack the relevant map in my toolbox, for I relish having a bird's-eye view of the terrain ahead once I have determined the direction the bees are heading when they fly home. Also, I enjoy knowing if there is a field or some other open place in the direction that the bees and I are headed. If so, then sometimes it makes sense to attempt a large jump of 400 or more yards down the beeline to this clearing and so quickly move in on the bees' forest home.

BIOLOGY BOX 2
Why Scent the Sugar-Syrup Bait with Anise Extract?

Anise extract (from the crushed seeds of anise, *Pimpinella anisum*) is not the only aromatic substance that can be used to scent the bee hunter's bait, but there is compelling evidence that anise extract is especially effective for this purpose.

In the summer of 1983, while conducting a study that looked at how honey bees adjust the strength of their recruitment of hive mates to a sugar-water feeder in relation to its energetic profitability (Seeley 1986), I learned that bees will recruit other foragers much more strongly to a feeder scented with anise than to one scented with peppermint, orange, or lemon, all else being equal. An undergraduate student, Ward Wheeler, and I were working with a honey bee colony that we had installed in a heavily forested region (the Yale Forest) in northeast Connecticut, to avoid interference from foragers coming from beekeepers' hives. There we set up two feeders at the same distance of 500 meters (1,640 feet) but in opposite directions (north vs. south) from our study colony, which was housed in a glass-walled observation hive mounted in a portable hut (see Seeley 1995, fig. 4.4). We then trained and labeled with paint marks—for individual identification—60 bees from this colony to forage at the two feeders; each feeder was visited by just 30 of the labeled bees (see fig. on following page).

On June 14, 1983, both feeders were filled with a sucrose solution of the same concentration (2.00 molar), and both sucrose solutions were given the same volume of scent (60 microliters of extract per

Experimental array and results of the test of whether sugar-syrup feeders scented with anise or peppermint differ in their effectiveness in attracting recruited bees. Thirty bees were trained to forage at each of two feeders, and these bees were labeled to identify them as the recruiters (*open circles*) to their feeder. All recruits (*filled circles*)—recognized as unlabeled bees—were captured upon arrival at the feeders. Whichever feeder was scented with the anise received many more recruits than did the feeder scented with peppermint.

liter of sucrose solution), but one feeder's food was scented with anise extract while the other feeder's food was scented with peppermint extract. Also, 3 milliliters of either anise extract or peppermint extract were placed in a vented reservoir beneath the slotted plate at the base of each feeder (see von Frisch 1967, fig. 21).

We tended the feeders steadily from 7:00 A.M. to 2:00 P.M., making hourly roll calls of the labeled bees visiting each feeder and steadily capturing all recruits arriving at each feeder. The recruits were recognized as such by their lack of paint marks and were captured in Ziploc bags shortly after they arrived at one of the feeders. From 7:00 A.M. to 10:00 A.M., the anise-scented feeder was in the south and the peppermint-scented feeder was in the north. Then from 10:00 A.M. to 11:00 A.M., both feeders were shut off (emptied), but some recruits continued to arrive and all were captured. Finally, from 11:00 A.M. to 2 P.M., we restored the feeders, but now the anise-scented one was in the north and the peppermint-scented one was in the south (to control for possible directional effects on recruitment), and we continued to take hourly roll calls of the labeled bees and capture the recruits (unlabeled bees) as they arrived at both feeders. We found that the rate of arrival of recruits was more than 10 times higher at the feeder scented with anise than at the feeder scented with peppermint, regardless of whether it was the north feeder or the south feeder that was filled with the anise-scented sugar solution. Similar results were found on the following two days when we tested anise vs. orange and anise vs. lemon.

I cannot say for sure why recruits arrived in greater numbers at whichever of the two feeders contained the anise-scented food, but I suspect that worker bees have an especially high sensitivity to the odor of anise. If so, then bees that follow dances announcing an anise-scented food source will find their recruitment target more easily than will bees that follow dances announcing a peppermint-scented (or orange-scented or lemon-scented) food source. Whatever the underlying mechanism, these experiments convinced me that scenting a food source with anise makes it especially attractive to recruits, so ever since then I have used only anise extract to scent my sugar syrup bait for bee hunting.

CHAPTER 3

Bee-Hunting Season

Where I live, in the Finger Lakes region of New York State, there is a rich tradition of hunting, especially deer hunting. So when I overhear folks at the gas station or grocery store talking about the hunting season I figure they are discussing when you can hunt white-tailed deer. In 2015, the deer-hunting season ran from October 1 to December 22. This period of big game hunting is a fairly dangerous time to be in the woods, and each year a dozen or so deer hunters in New York suffer a gunshot or arrow wound, so it is not surprising that the NYS Department of Environmental Conservation regulates deer hunting closely. For example, to be a legal deer hunter, you must be at least 16 years old. Moreover, you must be trained and certified in the safe use of your weapon—bow, crossbow, shotgun, or muzzleloader rifle—and you must keep careful track of the dates when you can legally use each type of weapon. You also must buy a hunting license, which specifies the number and sex of the deer that you can take. And eventually, once you are in the field, you must observe the many rules about the taking of deer: no using dogs, no setting salt licks, no hunting at night with a light, and no baiting.

Fortunately, the bee hunter—like the mushroom hunter and the wildflower enthusiast—can legally hunt his small game on any day of the year (fig. 3.1). The bee hunter can even do so on

FIG. 3.1. The start of a bee hunt on a September morning in an old meadow filled with goldenrod just outside the Arnot Forest. The top of Irish Hill is seen in the background.

the spur of the moment since he doesn't need to get a firearm certification or a hunting license. Furthermore, the bee hunter can use whatever hunting equipment and techniques he desires, such as baiting the bees with anise-scented sugar syrup. He must still, however, consider the time of year, for there is no point in going bee hunting when it is so cold that bees and flowers cannot be found. There is also the important fact that warm weather, by itself, does not guarantee good bee hunting.

WHEN DOES THE BEE-HUNTING SEASON START AND END?

Broadly speaking, the bee-hunting season runs from spring to fall. It starts in the spring as soon as the bees can fly out of their nests to collect food and water—that is, whenever the air temperature gets above about 55°F. In central New York, we sometimes begin to have days that get this warm in late March or early April. Eventually come the days when one is treated to the joyful sights of blooming crocuses, pussy willows, and red maple trees, all abuzz with bees collecting their first fresh food in many months. I have gone bee hunting at the start of the flowering season only once, and on that occasion the hunt unfolded quickly (see chapter 4), so I expect that early spring is generally a marvelous time to go bee hunting. This is because the bees are often nearly starving at the end of winter, in which case it is easy to get them excited about a sugar-syrup bait, even one that has a low sugar concentration.

I saw a striking demonstration of bees thrilling to low-quality food at the end of winter on a warm afternoon in late March 2007, when I was splitting blocks of freshly cut sugar maple into chunks of firewood: hundreds of honey bees mobbed my stack of newly split firewood to collect the sap. I know from tapping my sugar maple trees to make maple syrup that the sugar concentration of this sap was at most 3%. Like-

Fig. 3.2. Worker honey bee collecting nectar on a goldenrod (*Solidago* sp.) inflorescence.

wise, I have repeatedly seen crowds of bees at my bird feeders in the spring, collecting the dust from the cracked corn, presumably because these bees were desperate for some protein-rich food.

Bee hunting also works splendidly in the fall once frosts have destroyed most of the flowers, so your combs filled with sugar syrup are no longer competing with natural food sources for the attention of the bees. Recall, for example, how Thoreau was able to hunt bees on September 30, 1862, when a hard frost the previous night had withered the goldenrod flowers in the field beside Fair Haven Pond. On their way home, Thoreau and his companions managed to find a small patch of fresh goldenrod and purple aster flowers buzzing with bees (fig. 3.2) in a warm spot beside Walden Pond, and they had no difficulty getting these bees interested in their combs filled with diluted honey.

I had a similar experience on October 12, 2013, when I was retrieving bait hives—small hives mounted in trees to capture honey bee swarms—in the Arnot Forest. It was a delightfully sunny and warm afternoon, but for the past several nights there had been severe frosts. These had destroyed the goldenrod and purple aster flowers and so had eliminated the last of the bee's major food sources. Where two weeks before I had pushed my way through fields packed with lush, green, waist-high goldenrod plants whose yellow flowers glowed in the sunlight and teemed with bees, now I crunched through drab fields filled with the brown corpses of goldenrod plants that were, of course, devoid of bees. But when I approached a bait hive that I had installed in a white pine tree high up in the forest, I saw worker bees zipping to and from the hive. Had a swarm occupied the hive or, more likely given the season, had a robber bee from one of the wild colonies in the forest discovered some honey in the old combs that I had installed in this hive and recruited nest mates to help her exploit the find? Soon I had my ladder propped against the pine and had climbed up to the bait hive, so I could study the traffic of bees at its entrance.

Some bees were tumbling out, while others were feverishly pushing in; this certainly looked like the frenzied behavior of robber bees. And when I lifted the bait hive's lid and peered inside, my suspicion that this hive was the target of honey robbing was confirmed, for I found only about 100 bees inside the hive, and they were all either standing motionless sucking up honey on the one comb with some honey or running nervously over the other combs searching for honey. I also found the diagnostic sign of robbing: piles of wax flakes beneath the honeycomb where robbers had carelessly ripped the wax lids off the honey cells they were plundering. I'm confident that if I had wanted to establish a beeline then and there, I could have had hundreds of bees visiting my bee box within minutes of filling my little squares of comb with anise-scented sugar syrup.

THE STARTS AND ENDS OF HONEY FLOWS
ARE BEST FOR BEE HUNTING

While it is true that bees begin to work as soon as it gets warm in the spring and then continue to do so until there are severe frosts in the fall, there are times in between when the weather is fair and yet it is hard to find bees working on flowers. This happens because there are times when most of the foragers in a honey bee colony will stay at home, even if the weather is perfect for the bees. Why? It is because when a colony's foragers fly out of their home, they expose themselves to real predation risks and they incur high energy costs. For example, many foragers are caught by spiders, either in the webs of web-spinning spiders or on the flowers occupied by crab spiders. Many others are killed by aerial predators, such as dragonflies. Occasionally, while bee hunting, I will see one of my labeled bees take off ponderously from my comb, struggle to gain altitude with her abdomen bulging with sugar syrup, and then suddenly be snagged by an aerobatic dragonfly.

As for the high-energy costs of foraging, numerous studies of animal locomotion have shown that the means of flight used by bees—flapping of wings—is the most energetically demanding mode of locomotion. Indeed, the flight muscles of insects are, per unit of weight, among the most metabolically active of tissues. This means that a colony's foragers can achieve a net energetic gain from their labors outside the hive only when there are flowers offering the bees sizable payments in rich nectar for their pollination services. Sometimes there are massive blooms of flowers providing such payments, and if there is also sunny weather at these times, then colonies can fill their honeycombs with 10 or more pounds of honey in a day. Beekeepers call these times "honey flows."

The main honey flows for my location of central New York State come from black locust trees and dandelions in May;

sumac shrubs and raspberries in June; basswood trees, clover, and milkweed in July; and goldenrod and purple asters in August and September. I've kept a notebook of bee hunting for the past dozen years, and in it I've made a table of the dates of the findings of 21 bee trees. All are in July, August, and September, and all but one of the 21 bee hunts started with me catching bees on white clover, milkweed, or goldenrod. The one exception was on August 20, 2013, when I was doing a demonstration of bee hunting for the Catskill Mountain Beekeepers' Club (fig. 3.3). I met up with about 20 members of the club around 2:00 P.M. outside the Community Center in Acra, New York. A few minutes later we were all eagerly searching for foraging honey bees, for we were keen to get a beeline established to a wild colony. Most of us looked in the weedy places around the neatly mowed lawn of the Community Center, but one sharp-eyed member of the party spied honey bees busily foraging on a patch of wild thyme (*Thymus vulgaris*) plants growing in the tidy front lawn. We quickly scooped up some of these bees in our bee boxes, and within an hour we had established a roaring line pointing north that steered us to a colony living in the wall of an empty barn about a half mile away. Take-home lesson: when searching for bees to get a beeline started, look *all around* for flowers and inspect *all types* of flowers.

Generally speaking, the best times for bee hunting are when the bees are experiencing a definite honey flow, such as the milkweed flow or the goldenrod flow, for this means that it will not be hard to find bees on flowers. Bee hunting only works well, however, during *the start or the end of a honey flow*—that is, when nectar is available but is not super plentiful. The peak days of a honey flow are usually useless for a bee hunter because the rate at which a honey bee colony is taking in nectar has a strong effect on the motivation of its nectar foragers to recruit additional bees to their food sources. When a colony's rate of nectar intake is high, its nectar foragers are reluctant to activate additional bees and direct them to their forage sources

Fig. 3.3. Watching bees fly away from the bee box, to establish their beeline (flight route home).

(see biology box 3). This is true regardless of the source, be it a patch of flowers brimming with sweet nectar or a bee hunter's comb loaded with sugar syrup.

The bees' disinclination to bring nest mates to a comb filled with sugar syrup during the peak of a honey flow is a serious problem for the bee hunter. After all, once you have found bees on flowers, have caught a dozen or so bees in your bee box, have baited them with a comb filled with sugar syrup, and have released these bees to fly home, what you desire most keenly to happen next is for some of your bees to reappear quickly at your comb. Even more, you want your baited bees to bring lots of their sisters to your comb, so that you will have plenty of bees to observe flying home from where you are launching your hunt.

If the honey flow is just starting up or is winding down, then the bees that you've trapped in your bee box were probably experiencing only mediocre foraging success before you captured them. If so, then they are likely to be sufficiently impressed with your sugar syrup to want to return for more and to share with their nest mates the news of your wonderful free lunch. Indeed, if the bees are receiving only vanishingly small nectar rewards from the flowers, and the weather is delightful, then you could soon have dozens of bees mobbing your comb.

In early September 2011, I was treated to an unforgettable demonstration of the critical importance of low nectar availability to success in bee hunting. I was trying to line a bee tree from a clearing filled with goldenrod in the Arnot Forest, but I had made little progress. Only six individuals were visiting my syrup-filled comb, and their return trips were discouragingly sporadic. I concluded that the bees must be enjoying a whopper of a honey flow from the gleaming goldenrod flowers, and I was on the verge of calling an end to my hunt. But then tall, dark, booming storm clouds filled the sky, and soon a forceful thunderstorm was washing down the woods and old fields atop Irish Hill. It must have thoroughly rinsed the nectar from the goldenrod flowers, for once the storm had passed and I had restored my feeding station to give the hunt one more try, the bees were desperate for my sugar syrup, and they crowded on my feeder combs in the hundreds.

Usually, however, when a honey flow is in full swing it stays "on" for a week or so, and if you attempt to do a bee hunt during it you are apt to fail. Even if the bees that you've released from your bee box return to your feeder, they are unlikely to bring companions. With only a pitiful squad of halfhearted bees making occasional trips to your feeder, your enthusiasm for the day's adventure will soon fade. Sometimes things will go even more badly: your captured bees zoom away without looking back when you crack open your bee box, and that is the last you ever see of them. When the bees respond to your sugar syrup bait this indifferently, you had best pack up, go home, and wait a week or so for the honey flow to diminish. Remember, the most important quality of a bee hunter is patience.

BIOLOGY BOX 3
Why Don't Honey Bees Recruit During a Strong Honey Flow?

A striking feature of how honey bee colonies are organized to be efficient at making honey is the division of labor between the *foragers*, elderly bees that work outside the hive to gather nectar from flowers, and the *food storers*, middle-age bees that work inside the hive to process the freshly collected nectar, either distributing it among hungry nest mates or storing it in the honeycombs for future consumption (fig. A). This specialization by different-age bees on different parts of the honey-production process undoubtedly boosts the labor efficiency of a colony's workers. It means, for example, that once a forager bee has located a rich nectar source, she can concentrate on exploiting it before it runs dry rather than divide her efforts between the collecting and processing work. At the same time, however, this division of labor creates a problem of coordination within a colony because the rates of nectar collecting and processing must be kept in balance for the overall operation to proceed smoothly and efficiently. If the collecting rate exceeds the processing rate, then the foragers will experience unloading delays upon their return to the hive. Reciprocally, if the processing rate—or more precisely, the processing capacity—exceeds the collecting rate, then the food storers will experience delays in finding nectar foragers to unload. Either type of delay spells inefficiency.

Keeping the rates of nectar collecting and nectar processing matched is a difficult problem because colonies experience large and unpredictable swings from day to day in the availability of nectar, which is a function of the plants in bloom and the weather conditions. Because a colony tries to acquire as much nectar as possible, it will boost its nectar collection rate whenever the flowers start supplying more nectar during a honey flow. This means that a colony's nectar collection rate can change dramatically, even from one day to the next. For instance, on a day of cool, rainy weather, the nectar intake rate of a colony may be none at all, while a few days later, on a warm, sunny day

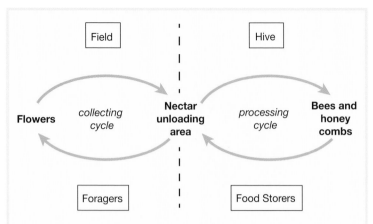

FIG. A. The two intersecting cycles of nectar collecting and nectar processing that make up the honey-production process. The collecting cycle operates mainly in the field as foragers gather nectar at flowers, bring it home, and then return to the flowers to gather more. The processing cycle takes place entirely in the nest as food storers unload nectar from newly returned nectar foragers in the unloading area (just inside the nest's entrance), transport the fresh nectar to other bees for immediate use or to honeycombs for storage, and then crawl back to the unloading area to repeat the cycle.

when the flowers are literally dripping with nectar, its nectar intake rate may surge to 10 or more pounds per day (see fig. 2.15 in Seeley 1995). Such large swings in a colony's nectar collection rate require, in turn, strong adjustments in the colony's nectar processing rate.

A honey bee colony is able to swiftly raise its nectar collection rate, or its nectar processing rate, whichever is needed. It can make these two adjustments because its nectar foragers can produce two different recruitment signals: the waggle dance and the tremble dance. The waggle dance (described in chapter 4) recruits more of a colony's elderly bees to the task of nectar collecting. It does so by giving them directions to rich sources of nectar outside the nest. The tremble dance (described in detail in Seeley 1995, pp. 162–173) recruits more of a colony's middle-age bees to the task of nectar processing. This task

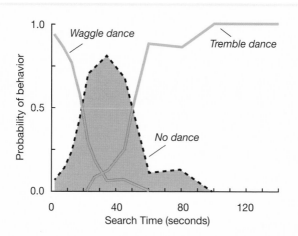

Fɪɢ. B. Dance behavior as a function of the in-hive search time for nectar foragers exploiting a rich source of nectar. A forager's "search time" is how long she needs to search inside the nest to find a food-storer bee that will take her load of freshly collected nectar. If a bee experiences a short (< 20 sec) search time, then she is likely to perform a waggle dance, but if her search is long (> 50 sec), then she will perform a tremble dance instead.

takes place inside the nest. A food-storer bee meets a returning nectar forager just inside the nest's entrance, extends her tongue to the mouth of the forager, and imbibes the droplet of nectar that the forager regurgitates. Next, the food storer transports the fresh nectar to a location deep inside the hive, where she either feeds it to hungry bees or deposits it in the honeycombs. Finally, she crawls back to a spot near the nest entrance to unload another nectar forager.

When a nectar forager returns to her colony's nest from a patch of flowers brimming with nectar—or from a bee hunter's comb stocked with sugar syrup—she must decide whether to perform a waggle dance, a tremble dance, or neither. We now know that she makes this decision based on how long she has to search inside the nest to find a food storer (fig. B). If the nectar forager needs to search for only 20 seconds or less, which indicates that plenty of bees are already functioning as food storers, then she will perform a waggle dance to direct more foragers to the rich food source she has encountered. But if she must search for 50 seconds or more, which indicates that more food storers are needed in the hive (just as a long line at a bank indicates that more tellers are needed), then she will perform a tremble dance to stimulate more middle-age bees to function as food storers. And if she searches for 20–50 seconds, which indicates that the "honey factory" is running smoothly, with its rate of nectar collecting and nectar processing efficiently balanced, then she will perform neither waggle dance nor tremble dance.

The fact that nectar foragers will switch from waggle dancing to no dancing, or even to tremble dancing, when they experience lengthy delays upon return to the nest, explains why bee hunting fails during a strong honey flow. When nectar is super abundant, the nectar foragers can load up quickly at the flowers and so will return frequently to the nest. A brisk traffic of returning nectar foragers will give rise to long unloading delays (more than 20 seconds), which in turn will inhibit these bees from performing waggle dances. Therefore, during a powerful honey flow, even if some bees continue to exploit a bee hunter's rich bait of a comb filled with sugar syrup, they are not likely to perform the waggle dances that are needed to give the bee hunter a good bevy of bees at his comb.

 CHAPTER 4

Establishing a Beeline

Let us assume that it is a warm day, wildflowers are in bloom, and you are keen to be outdoors in sunshine and fresh air, hunting for a wild colony of honey bees. We will also assume that you have built a bee box and assembled the rest of your bee hunter's toolkit, and that you have decided where you would like to start your first hunt. This could be in your back-yard, on a friend's land, or at a state park or nature preserve. It could even be a public space in the heart of a city. Central Park in New York City, for example, would be perfect. I suspect that several wild colonies of honey bees live in the wooded portions of this 843-acre masterpiece of urban landscape architecture.

Some years ago, I had fun conducting an urban bee hunt in the center of Cambridge, Massachusetts. This hunt took place on a bright and warm Sunday morning in April 1979; it grew out of something I had seen two days before while walking through Harvard Yard, the oldest part of the Harvard University campus and a handsome green space in densely built-up Cambridge (fig. 4.1). I had noticed that a bed of crocuses in front of Memorial Church was buzzing with honey bees gathering pollen from the purple, yellow, and white blossoms. While enjoying the sight of these bees, I started wondering what their home address might be. I had previously found three colonies of honey bees living in hives in the yard of a house next to the Harvard Divinity School, about a half mile from the Yard. This little apiary was well hidden behind a six-foot fence of vertical boards that enclosed the house's yard, but the traffic of worker bees zooming over the fence had attracted my attention. By peering through a few slits between the fence boards, I could just see the hives. I had also spotted a wild colony of honey bees living in a sugar maple tree along Oxford Street, in front of the Museum of Comparative Zoology, about a quarter mile from the Yard. So I wasn't at all surprised to find honey bees on the crocuses beside the Harvard Memorial Church, but I was curious about where they were coming from, and I was eager to test my skills as a bee hunter in this city setting.

I decided to act on my curiosity and conduct a bee hunt that would start at the crocuses in Harvard Yard. I also decided that this bee hunt would be in memory of Professor George Harold Edgell, who must have walked through this place often, coming and going from the nearby Fogg Museum of Art. If I had known then that Henry David Thoreau had also been an avid bee hunter, then I would have conducted the hunt in his memory too, for Thoreau certainly walked around here when he was a student at Harvard College (1833–1837) and lived in Hollis Hall, which is in the Yard. I started my hunt by capturing bees off the crocuses along the front of Memorial Church, where I had first noticed them. In less than an hour, using the methods I'm about to describe, I had a strong line of bees running to the southeast. Although I certainly did not know it when I began this hunt, the home of these bees was right there in Harvard Yard. Indeed, it was in sight of my starting point! Searching in the direction the bees flew off when they headed home, I soon discovered their residence: a crevice above the massive doors on the west end of stately Emerson Hall, only 120 yards from the bed of crocuses.

Coincidentally, two years before, I had shaken a swarm of bees from the branches of a tree beside the Harvard Faculty

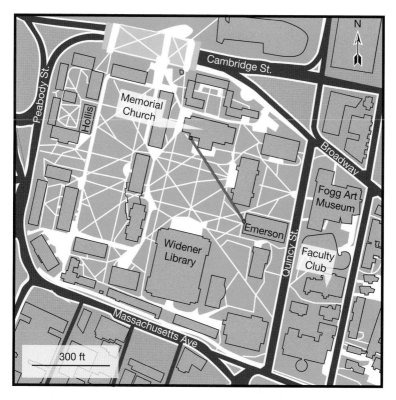

FIG. 4.1. Map of Harvard Yard. Red line denotes the path of a bee hunt that began at a bed of crocuses in front of Memorial Church and ended at a colony living in the west wall of Emerson Hall, just 120 yards away.

Club, which is located just outside Harvard Yard. This tree is only 100 yards from the west end of Emerson Hall. Of course, while collecting this swarm, I had wondered where these bees had come from. Now I had a likely solution to this mystery.

CAPTURING BEES FROM FLOWERS USING THE BEE BOX

Whatever the location you have chosen for launching your hunt, your first task upon arriving there is to find a patch of

FIG. 4.2. Worker honey bee collecting pollen from pussy willow (*Salix discolor*).

flowers with honey bees. Ideally, it will be a sunny spot chock-full of bees, but this is not essential. I have established many beelines from flower patches where I had to hunt for 5 or 10 minutes to find just one honey bee. Recall that this is what Kirk Visscher and I experienced on our first bee hunt, described in chapter 1, when we looked for foraging honey bees in the Arnot Forest and had to search for nearly half an hour before finding one on a multiflora rose bush in bloom. Of course, before you go looking for honey bees, you will want to be sure that you know what a worker honey bee looks like (fig. 4.2). Honey bees are generally leather-colored and are much smaller and sleeker than the big, fuzzy, black-and-yellow bumble bees that many people mistake for honey bees.

Once you have spied a honey bee on a flower, your next job is to catch her (fig. 4.3). This involves opening fully the door of your bee box (while holding the divider in the middle closed tightly) and moving it slowly up to the bee. Then, when you have positioned the mouth of the box within an inch or two of the bee, you quickly push it forward to get the bee inside the box while she is still on the flower. In the next instant, before your bee can fly off, you slap the door shut. Performing this maneuver deftly is not difficult, but it will take some practice. And it is definitely easier with the flowers of some plant species than others. The best are those with either high flowers, such as goldenrod (*Solidago canadensis*) and milkweed (*Asclepias syriaca*), or small flowers at the ends of slender stems, such as dandelions (*Taraxacum officinale*), white clover (*Melilotus alba*), knapweed (*Centaura* spp.), Queen Anne's lace (*Daucus carota*), and chicory (*Cichorium intybus*). With practice, you will succeed in capturing maybe eight out of ten bees on the first try. Fortunately, a bee that evades your first crack at catching her usually will not fly away and instead will tend to alight on a nearby flower to continue foraging, giving you another shot at snagging her.

It is essential that bees on flowers be captured. In my experience, there is no point in simply setting out a syrup-filled comb and hoping that bees working on flowers at the site will abandon them and switch to exploiting your comb, even if you have anointed it with the alluring aroma of anise. It is true that when flowers are conspicuously scarce, such as after several hard

frosts in autumn, foraging bees will be attracted to an aromatic, syrup-filled comb. But this is the exception, and even in such an extreme starvation situation you may have to wait hours for a forager bee to discover your free lunch. Almost always, the bees will be searching elsewhere or resting at home.

In his book *The Bee Hunter*, George H. Edgell relates how just once, in more than 50 years of bee hunting, did he start a line by accident. It was in the fall, after the frosts had destroyed nearly all the flowers. He hiked to a sheltered clearing where he hoped he might find a flower or two, and there he poked about searching for a bee to capture in his bee box. Meanwhile, he left a spare bee box containing an empty comb sitting open on a boulder. After searching for a bee without success for 15 or 20 minutes, he returned to gather up his equipment and found one honey bee circling the empty comb, no doubt attracted by the scent of the comb and anise. He then filled his dropper with syrup and managed to drip some onto the comb without frightening the bee. She landed, loaded up, flew off, and soon returned. Her colleagues started joining her a few minutes later.

The only times that I have established beelines without capturing bees off flowers with my bee box are the couple of times when I was out bee hunting on hot days in August and caught sight of bees loading water at the edge of a pond, stream, or other wet spot (fig. 4.4). Honey bees will cool their nest on hot days by spreading water on the combs and drawing air through the nest by fanning their wings. This produces strong evaporative cooling. To get the water collectors hooked on my sugar syrup, I use the dropper bottle in my toolkit to drip some sugar

Fig. 4.3. Capturing a bee off a flower using a bee box. *Top:* Reaching toward the bee while holding the bee box's door open. *Middle:* With the bee box's door held shut, the middle divider is raised and the shutter that covers the window in the rear compartment is opened, to lure the bee to the back of the bee box. *Bottom:* The middle divider is lowered, to confine the captured bees in the rear compartment.

Fig. 4.4. Worker honey bee collecting water from a pond surface while standing on the floating fronds of common duckweed (*Lemna minor*).

syrup onto the wet surfaces on which the bees are standing. Once a bee begins to probe the sugary spot with her tongue, indicating that she has detected its intoxicating sweetness, I gently place a square of comb beside her, being careful to not bump her. Then I gingerly squirt a drop of syrup on the side of the comb so it dribbles down beside the bee. Usually, this induces her to climb onto the comb and discover its amazing supply of first-class forage.

We will assume, though, that you have not stumbled upon a water collector and coaxed her to climb onto your syrup-filled comb, and that instead you have caught a forager in the front chamber of your bee box. The next step is to introduce her to a syrup-filled comb. To do this, first raise the shutter covering the window in your bee box and then raise the divider in the middle of the box, making sure to not let it (and the bee) come all the way out of the box. Seeing the light, the trapped bee will run to the back chamber, trying to escape. When you see through the window that she is in the rear compartment, lower the divider (to lock her in) and then lower the shutter (to darken the rear compartment). Now you can safely open the front door to catch another bee. You can establish a beeline with just one bee, but you will have a higher probability of success if you can capture a half dozen or more in your bee box. After imprisoning the first bee, I try for 5 or 10 minutes to catch more bees before I take the next step of releasing the captured bee(s) to start the beeline.

CHOOSING THE RELEASE SITE

Before you set your prisoners free, you must make an important decision: where exactly to position your bee box when you release the bees. The ideal spot is one that offers a clear field of view for 100 feet or more in all directions, so that after you free the bees you will be able to watch them for a long ways as they fly off. This will enable you to get good readings of their vanishing bearings once they become regular customers at your little filling station. The center of a broad meadow is ideal, but usually one makes do with a less spacious spot, such as a lawn or a clearing alongside a road. If you have brought along a stand for the bee box, set it where you will have the longest lines of sight, making sure it is on even ground so the stand won't wobble.

Now set your bee box on the stand (fig. 4.5). Fill one of your comb squares with syrup, place it in the front compartment, shut the door, and raise the divider, propping it at an angle so it stays up. Then spread your handkerchief-size piece of light-proof cloth over the bee box and leave everything undisturbed. (Note: without the cloth over the box, the imprisoned bees will struggle to escape at the cracks where sunlight is leaking in, and so will be less apt to bump into your syrup-filled comb.) After five minutes, uncover the bee box and gently open its door. Most of the bees will have discovered the syrup and

FIG. 4.5. Introducing bees to sugar-syrup bait. *Top:* The door and shutter of the bee box are closed and the divider is raised, so that bees confined in the rear compartment can find the syrup-filled comb in the front compartment. *Bottom:* Bee box as above, but now darkened with an opaque cloth.

loaded up on it in the darkness, though one or two may be still stuffing themselves. Some that are fully loaded will be grooming to remove tiny smears of syrup on their wings or antennae. In a minute or two, though, all the bees that have loaded up will fly off. When they have all departed, take the syrup-filled comb from your box, set it on the stand so that it will be available to the bees you've just released upon their return, and repeat the whole process of capturing bees off flowers, giving them access to syrup, and finally letting them go. By repeating this capture-and-release process, you multiply your chances of success in establishing a beeline.

WATCHING THE BEES FLY AWAY

When a bee departs your feeding station for the first time, she will first turn back and look at the site and then will fly off slowly in gradually expanding circles and figure eights. She performs these stereotyped flight maneuvers to memorize the look of the bee box and the arrangement of the landmarks around it so that when she comes back she will be able to find her way straight to your syrup-filled comb. You, of course, are itching to see the direction in which she disappears so you can learn the direction to her home. As she circles, you twist and turn, perhaps pointing at the bee to help keep your eyes locked on her. Sometimes when she flies between you and the sun she is lost from view. Other times she disappears when she flies against a background of dark trees. You will have performed admirably if you can establish from a bee's first flight home whether she has headed more north than south, or more east than west. Most bees will need to make a half dozen or so trips to and from your bee box before they will be sufficiently familiar with the box and the landmarks nearby to fly off in a straight course home—a true beeline.

If the conditions are right, meaning you are not trying to hunt during the peak of a honey flow (see biology box 3 in chapter 3), then at least half of the bees that you released will return for another load of your delicious sugar syrup. They may even recruit nest mates to your marvelous food source by performing waggle dances at home to guide others to it. If you are hunting when the flowers are secreting only meager amounts of nectar, so that your syrup-filled comb has little competition for the bees' attention, then this recruitment can happen quickly. Within an hour of the release of your bees, you might have a dozen or more bees simultaneously tanking up at your comb, as was the case during my bee hunt in Harvard Yard. If, however, you are hunting during the peak of a strong honey flow, or if you have happened to catch bees whose home is far away, then you might find that none of the released bees come back, or that these bees will repeatedly collect loads of syrup but will bring no nest mates. This is frustrating, and the best course of action is to wait a week and try again, or to start over in another location, which might be closer to a bee tree.

We will assume, though, that the conditions are right on your first day of bee hunting. We will also assume that the home of the bees you caught is not miles away, so that one of them returns less than 10 minutes after you released your bees. Now comes one of the most exciting moments in a hunt, for you want desperately for this returning bee to land and load, but she is exercising due caution and does not settle immediately. Normally, the only time that a foraging bee imbibes syrupy food straight from a beeswax comb is when she is stealing honey from another colony. This is dangerous work! Robber bees are often caught and killed by the guard bees of the colony being raided. So, unsurprisingly, your bee behaves nervously, at first circling the comb without landing, then darting out of sight, but eventually returning and approaching the comb in narrowing circles until she hovers just above the comb, her little wings buzzing shrilly. You sit motionless, with bated breath, lest you spook her. At last, she touches down, her buzzing ceases, and she inserts her tongue in the sugar syrup (fig. 4.6). She is calmly

FIG. 4.6. Worker bee with tongue extended, drinking sugar syrup from a square of beeswax comb.

loading! From now on, she is your bee. That is, so long as you keep offering her rich food, she will keep visiting your comb. You have started your beeline.

Within a minute or two, your little helper will finish her loading and depart. Again, you twist and turn, and again you roll your head around trying to keep her in sight as she departs in gradually widening circles, for she is still learning how to find her way home. As before, you will probably lose her against the bright sun or dark trees before she shoots off in a straight line, but you will have seen her circling mostly in one general direction, so you can adjust your position to see her better the next time. Soon, other bees arrive. Most are individuals that you caught on flowers and released, but some are newcomers that followed the waggle dances performed by your original bees inside their bee-tree home. By following these dances, these bees learned the location and scent of your sumptuous free lunch.

Now that you have numerous bees visiting your comb, it is time to label about 10 of them with paint marks for individual recognition. This will make the bees' affairs personal and thus vastly more interesting. Bees do not like to be painted, but if you touch a bee deftly and gently with a tiny camel's hair brush or the wet nib of a paint pen, then she will receive your decoration without being excessively disturbed. It is easiest to daub the paint on top of a bee's thorax, which is the middle part of her body. The target for your paint mark is the fuzzy spot between the wings, but take care to apply just a pinhead-size dot of paint (fig. 4.7). Do your best to avoid getting paint on either the bees' wings or the hinges at the bases of the wings, where they attach to the thorax, for even the tiniest speck of paint on a wing tip or in a wing hinge will elicit prolonged grooming. A worker bee works hard to keep her flight machinery in good repair. A skilled bee hunter can also daub a paint mark on the top of a bee's abdomen (the rear section of her body), but this requires that the bee has her wings spread widely, not folded over her abdomen. Sometimes you will spy a bee that is sucking from a cell on the edge of the comb and has let her abdomen droop so that its dorsal (top) surface is away from her wings; such bees are easily given abdomen paint marks without muss or fuss.

GETTING GOOD SIGHTINGS

By now an hour or two has passed since you released the first cohort of bees from your bee box, the comb has 10 or more bees on it at times, and arrivals and departures are frequent. Also, many of your bees may have come and gone 10 or more times and become accustomed to the site, so as soon as they are loaded they launch into flight and fly straight off. Unless an

FIG. 4.7. Labeling a bee with a paint dot on the thorax while she stands, unmoving, drinking sugar syrup from a feeder comb.

outbound bee zooms toward the sun, you will have a good chance of tracking her for 50 or more yards before she disappears. Each time you manage to make one of these long-range sightings, make a mental note of a landmark in the direction in which you last saw the bee, and then get a reading of the bee's vanishing bearing using your compass. These compass readings provide excellent information, and I like to record them in a table in my notebook; doing so helps me see which direction is emerging as the main route home for the bees. Of course, there will be scatter in your data (fig. 4.8). Different bees can adopt different routes home, even if they come from the same nest. If, for example, there is a tall pine 50 yards away in the general direction of the bees' home, then some bees might fly around it to the left, while others will bypass it to the right, and a few will choose to fly up and over it. Using your notes of the bees' departure directions, you can calculate the average of all these

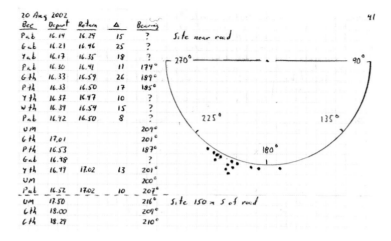

Bee	Depart	Return	Δ	Bearing
P ᵤ b	16.14	16.29	15	?
6 ᵤ b	16.21	16.46	25	?
7 ᵤ b	16.17	16.35	18	?
P ᵤ b	16.30	16.41	11	174°
6 th	16.33	16.59	26	189°
P th	16.33	16.50	17	185°
Y th	16.37	16.47	10	?
w th	16.39	16.54	15	?
P ᵤ b	16.42	16.50	8	?
U m				209°
6 th	17.01			201°
P th	16.53			187°
6 ᵤ b	16.78			?
Y th	16.77	17.02	13	201°
U m				200°
P ᵤ b	16.52	17.02	10	207°
U m	17.50			216°
L th	18.00			209°
L th	18.29			210°

Fig. 4.8. Page in bee-hunting notebook showing data on away times and departure bearings. There is much scatter in the directional data, but the general pattern is clear: the direction to the bees' home is about 200°.

directions and get a best estimate of the true direction to the bees' home.

Occasionally, however, the average value of your bees' vanishing bearings is misleading. On August 29, 1978, for instance, I established a beeline at a site along the dirt road that runs east–west in the northern part of the Arnot Forest. I did so using foragers I found working a patch of milkweed plants (*Asclepias syriaca*). Incidentally, I also found in this milkweed patch a shiny, almost turquoise blue chrysalis of a monarch butterfly (*Danaus plexipuss*). This is the only monarch chrysalis that I've found in the wild, and it is not by chance that I discovered it while bee hunting. You see, when I'm out hunting the bees, I'm working slowly and directing my full attention to the task at hand. After all, the key ingredients of a successful bee hunt are spying and capturing bees on flowers, labeling individuals and watching their homeward flights, and finally catching sight of bees diving into a knothole or crevice in a tree.

You will accomplish these things best if you can focus your attention—especially your visual attention—on the flowers before you, the bees visiting your feeding station, the trees along their flight route, the weather signs, the lay of the land, and whatever else that will help you track down your wild quarry. Bee hunting has sharpened my skills in observing nature closely, and I'm sure it will do the same for you.

Fortunately, the bees I found and captured in the milkweed patch quickly accepted my comb as a desirable food source. When they finished loading and took off, most flew straight down the road, which ran west at 260°. I eventually learned, however, that 294°, not 260°, was the true direction from this milkweed patch to the bees' home, a sugar maple tree three-quarters of a mile away. So the destination of these departing bees was fully 34° north of the average of their vanishing bearings. It seems that these syrup-stuffed bees had found it easiest to fly home by first heading west down the open flight path provided by the road and then banking right, on a more northerly course, after they had gained enough altitude to clear the stand of tall spruce trees running along the north side of the road.

I had another experience of being strongly misled by the bees when I was hunting in the Arnot Forest, though what happened this second time was curiously different from the case just described. On September 20, 2002, I started a hunt in a clearing beside the McClary Road, a dirt track that has been abandoned for probably 60 years now. It runs southwest for nearly 2 miles from atop Irish Hill down past the stone-walled cellar holes of several farm houses, then along the rim of a shady, hemlock-lined gorge that holds the lovely, but rarely visited Opalescent Falls; it eventually emerges from the Arnot Forest when it reaches the floor of the valley through which the Cayuta Creek flows. I had no trouble finding honey bees foraging on the glowing blossoms of goldenrod (*Solidago canadensis*) growing in an abandoned field (site 10 in fig. 6.2), and soon

I was watching bees leaving my feeder comb in two general directions, 158° and 198°, hence a bit east and a bit west of south. Over this day and the next two, I worked my way down the beeline that pointed southeast in a series of moves (as explained in chapter 6) and eventually located a bee tree 0.54 miles from my starting point (bee tree G in fig. 6.2). A few days later, on September 24 and 25, I conducted another successful bee hunt in the same corner of the Arnot Forest (from site 11 to bee tree H, shown in fig. 6.2), and in doing so I reconnected with four of the labeled bees that had flown off at 198° from site 10 back on September 20. Now that I knew the exact home address of these four bees, I could see that when they had flown home on September 20 (from site 10) they had piloted themselves along a curved flight path, not a proverbial beeline. Their curiously curvilinear track home took them around, not over, a protruding hillside, so they had enjoyed homeward flights that were entirely downhill. If these bees had flown straight back to their nest, they would have made shorter flights home but each one would have required a 100-foot climb over the intervening hill. This would have been energetically costly for these bees, whose abdomens were bulging with sugar syrup, so probably it was more fuel-efficient to use the longer, but continuously downward, flight route. The important point for now, though, is that if these bees had flown straight home, their average vanishing bearing would have been nearly due south, 182°, instead of the 198° that I had measured.

These two instances in which the bees adopted indirect, but sensible, flight routes home illustrate something about honey bees that the aspiring bee hunter should know and respect and enjoy: these little wonders are probably *the most intelligent and behaviorally versatile insects in the world*. We see this in the way a foraging bee exquisitely adjusts her behavior to adapt it to the particulars of her current situation, including the weather conditions, the time of day, the nutritional needs of the colony, the profitability of the flowers, and the obstacles she encounters. It is because these brainy insects are endowed with so much behavioral adaptability that the bee hunter cannot have blind confidence in his estimate of the direction to his bees' home based on their vanishing bearings. As we have just seen, sometimes the bees use curvaceous routes home to dodge obstacles. The bees' behavioral adaptability is also the reason—as we shall see in the next chapter—the bee hunter must carefully interpret his measurements of how long the bees are gone from his feeder when he estimates the distance to their home. I hasten to emphasize that these two wrinkles in interpreting the bees' behavior are actually positive features of bee hunting. After all, it is the complexity of the bees' behavior, together with the uniqueness of each hunt's location, which makes bee hunting a sport of such infinite variety, intellectual challenge, and fun.

BIOLOGY BOX 4
How Do Honey Bees Recruit Nest Mates to Food Sources?

When a worker honey bee finds a rich source of nectar or pollen, she is able to recruit nest mates to it and thereby strengthen her colony's exploitation of this attractive food source. The principal mechanism of this recruitment process is the waggle dance, a unique behavior in which a worker bee, deep inside her colony's nest, performs a miniaturized reenactment of her recent journey to the rich food source. This is usually a flower patch, but it can also be a bee hunter's syrup-filled comb. Bees following the dancer learn the distance to the food source, the direction in which it lies, and its scent. They then use this information to steer their flights to the specified location.

To see how bees communicate using waggle dances, let us follow the behavior of a bee upon her return from a highly profitable food source. We will suppose that the bees' find is a patch of flowers brimming with nectar located a moderate distance from her nest, say 1,000 meters (about 1,100 yards, or 0.6 miles), and along a line 40° to the right of an imaginary line pointing in the direction of the sun (see fig.). Excited by her foraging success, the bee scrambles inside her colony's nest cavity and immediately crawls onto one of the vertical combs. Here, amid a massed throng of her fellow worker bees (all of them daughters of the colony's queen), she performs her recruitment dance. This involves running through a small figure-eight pattern: a waggle run followed by a turn to the right to circle back to the starting point, another waggle run, followed by a turn and circle to the left, and so on in a regular alternation between right and left turns after waggle runs. The waggle run portion of the dance is the most striking and informative part of the bee's performance, and the dancing bee gives it special emphasis both by the vigorous waggling of the body—the lateral shaking of the body, with sideways deflections greatest at the tip of the abdomen and least at the head—and by the up-and-down vibrating of the wings at approximately 260 hertz (Hz; cycles per second).

Usually several bees will trip along behind a dancer, their antennae always extended toward her. These followers detect the dance sounds with their antennae. The outermost portions of a worker bee's two antennae have a resonant frequency of about 260–280 Hz, matching the vibration frequency of the wing vibrations. Moreover, the vibration detector at the base of each antenna (the Johnston's organ) is maximally sensitive to vibrations in the 200- to 350-Hz range.

The direction and duration of each waggle run are tightly correlated with the direction of and distance to the food source being advertised by the dancing bee. A flower patch, or a bee hunter's comb, that is located directly in line with the sun is represented by waggle runs produced while the bee walks straight upward on the vertical comb, and any angle to the right or left of the sun is coded by a corresponding angle to the right or left of the upward direction. In the example illustrated in the figure, the flower patch lies 40° to the right of the sun's direction, and the waggle run is correspondingly aimed at the angle of 40° to the right of straight up. The distance between the

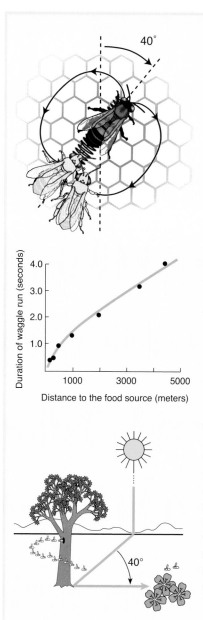

How a dancing bee provides information to other bees about the distance and direction to a rich food source. **Top:** The movement pattern of a worker bee performing a waggle dance on the vertical surface of a comb inside her colony's nest. The dancing bee walks straight ahead on the comb, waggling her body from side to side, then she stops the "waggle run" and turns left or right to make a semicircular "return run" back to her starting point. This bee may perform dozens of these dance circuits consecutively. Two follower bees are acquiring the dancing bee's information. **Middle:** How distance information is expressed; the distance to the food source is indicated by the duration of the waggle runs. **Bottom:** How direction information is expressed; outside the hive, the dancing bee notes the angle of her outbound flight relative to the sun's direction, and then inside the hive she orients her waggle runs at the same angle relative to straight up on the comb.

nest and the food source is indicated by the duration of the waggle run. The farther the food source, the longer the waggle run in each circuit of the dance, with a rate of increase of about 75 milliseconds of waggle run duration per 100 meters (110 yards) of flight distance. Workers can detect the buzzing sound produced during a waggle run, so presumably the dance followers perceive the duration of a dancer's waggle runs by sensing the duration of the sound associated with each waggle run.

Besides providing information about direction and distance, a dancing bee provides information about the odor of the food at her forage site. This scent is carried back to the nest in the waxy layer ("cuticle") that covers her body, but often a stronger source of the scent is the food she brings home—the loads of pollen on her hind legs or the nectar (or sugar syrup) she regurgitates to the dance followers. A bee that is following a waggle dancer rapidly memorizes the scent wafting off the dancing bee. Recruited bees then draw upon their memory of the advertised food source's scent to help them pinpoint its location after using the dance's direction and distance information to arrive in the general vicinity. This is why it is critically important that the bee hunter add a droplet of distinguishing anise extract to the sugar syrup that functions as his bait.

 CHAPTER 5

Timing Bees to Estimate Distance to Home

Now that you have a squadron of bees flying confidently to and from your comb, many of them labeled with eye-catching paint marks for individual identification, you are ready to start recording the departure and return times of these bees to measure how long they are away from your comb when they make trips back to their secret residence. We will see that these "away times," if used judiciously, can give you a good estimate of the distance from your comb to their home. You will want to get a sense of this distance as quickly as possible after you have established a beeline, because the distance varies greatly among hunts and it strongly affects the difficulty of solving the mystery of your bees' home address. Reviewing the records in my notebook for bee hunting, I see that for my 21 successful hunts made over the last 15 years—18 in the Arnot Forest, south of Ithaca, New York; 2 in the Powdermill Nature Reserve, east of Pittsburgh, Pennsylvania; and 1 in the Catskill Mountains, near Acra, New York—the distance from my starting point to the bees' home varied widely from hunt to hunt. The shortest was 160 feet and the longest was 1.2 miles. The average was 0.46 miles. The notebook also contains records of another 8 hunts that I started but aborted, usually because the distance to the bees' dwelling place was so great that the captured bees would not bring others to my feeder comb. Twice, though, I had a bad day and failed to find the bees' hidden nest, even though I knew I had gotten close to it. Failures in bee hunting certainly do happen, despite the bee hunter's strongest efforts and most fervent desires, and will be discussed in chapter 7.

My rough guidelines relating how long the bees are gone from your comb and how far it is to their home are as follows. If the bees are gone only 2–4 minutes, the bee tree is very close, possibly within sight. If the bees are gone 5–9 minutes, the bee tree is less than a mile away. If the bees are gone 10–15 minutes, the bee tree is far away but still findable. If the bees are gone more than 15 minutes, then things are pretty hopeless because the bee tree is probably more than a mile and a half away. Bees are not likely to recruit nest mates over such a great distance, so probably you will never get a heavy traffic of bees at your comb. When this happens, the best plan of action is to abandon the stand, move about a mile in the direction your bees have flown, and restart the whole process by capturing more bees off flowers at the new location.

MEASURING AWAY TIMES

In principle, it is simplicity itself to figure out how much time a bee spends away from your comb on a trip home: you record her departure and return times and then figure the difference between them. It is easy to get good readings of your leaving bees' departure times because these bees show distinctive behaviors that reveal when they are about to leave for home. In particular, when a bee finishes drinking from your comb, she will fold her tongue into its rest position beneath her head, and she might groom her wings and antennae in preparation for her flight. And when she finally takes flight, she is apt to buzz off slowly, for she will be laden with a syrupy payload that has nearly doubled her flight weight. It is not so easy to get accurate readings of your returning bees' arrival times, however, because

Fig. 5.1. Taking notes on when individually identifiable bees leave the feeder comb and when they return there, to measure their away times. Properly interpreted, these times yield a reliable estimate of the distance to the bees' nest.

these bees do not alert you to their impending arrivals. Also, because they are flying without cargo, they can zip in quickly and land unnoticed. Because it is so easy to miss the arrivals of returning bees, you must train yourself to keep a sharp lookout for the reappearances of bees for whom you have a departure time but for whom you still need a return time. Keeping track of which members of your bee team are due to arrive will give your memory a good workout. Even if blessed with a superb memory, you will need a notebook for recording your bees' departure and arrival times and the values of how long your bees were away from your feeding station (fig. 5.1).

CONVERTING AWAY TIMES TO ESTIMATES OF DISTANCE

What is even trickier than acquiring good data on your bees' away times is interpreting these data to get an accurate estimate of the distance to their dwelling place. You are probably dying to know whether your bees are living in the woods beside the field where you've begun your hunt or are dwelling high up on the forested hillside that rises in the distance. To begin to understand how to interpret your away-time measurements, note that it takes your labeled bees one to two minutes to load up at your comb; this is a good guide to the *minimum* time it takes a bee to offload her "nectar" when she gets back to her nest. A returning forager also needs a bit of time—generally half a minute at least—to crawl in the nest entrance and find one of the middle-age bees that are functioning as her colony's food storers. At a minimum, your labeled bees will spend approximately two minutes at home between trips to your comb.

Sometimes, however, a forager will spend more than two minutes at home between foraging trips, a fact that is important to keep in mind when interpreting your data on away times. To see how these longer stays at home can arise, let's follow a bee that has returned home from your filling station, has crawled onto a comb just inside the nest entrance, and has found a nest mate eager to receive her load of nectar. With her tongue retracted, the forager spreads her mandibles and begins to regurgitate the droplet of sweet liquid that she has airlifted home. Simultaneously, the food-storer bee extends her tongue and steadily drinks in the rich food (fig. 5.2). This unloading process usually lasts only a minute at most, and when it is completed the food-storer bee starts to "ripen" the fresh sugar solution into honey. She adds the enzyme invertase, to cleave the sucrose molecules into the more soluble sugars fructose and glucose, as well as the enzyme glucose oxidase, to generate hydrogen peroxide that will protect the honey from spoilage. The food-storer bee then climbs up to the honey comb region in the top of the nest and deposits her load in a beeswax cell that, when full, will be sealed with a beeswax lid to reduce moisture absorption by the honey. Fully ripened honey contains only 14–18% water and is hygroscopic (that is, it will absorb moisture from the air).

Fɪɢ. 5.2. A nectar forager (right), having returned to the nest, regurgitates her load of nectar (or sugar syrup) to a food-storer bee (left) that has inserted her tongue between the mouthparts of the forager.

While the food-storer bee is busy converting your sugar syrup into honey, the forager bee that brought home the goodies from your comb might be dashing pell-mell to the nest entrance to get back to your comb without delay. If so, then she will spend only about two minutes inside the nest. It is also possible, however, that this forager is so excited by your profuse supply of sugar syrup that she won't race back out to you and instead will busy herself in the nest for several minutes performing a waggle dance (fig. 5.3). This miniaturized reenactment of her flight to your comb will amplify the colony's exploitation of your food bonanza by directing other foragers to the site. It is also possible that your forager bee will opt to take a rest break or groom herself (or both) before embarking on her next collecting trip. Of course, if she chooses to spend time dancing, resting, or grooming, then she will spend more than two minutes in the nest between excursions to your feeder comb.

Because only some of your bees will spend the least possible time in the nest between trips to your feeder, you will want to base your estimate of the distance to the bees' home on only

Fig. 5.3. A worker bee is performing a waggle dance that the bees behind her are following.

the shortest away times that you record. Let's look at an example of how this is done by reviewing the data I collected when I started a bee hunt on July 30, 2011. It was a sunny Saturday, and I had left my house around 7:00 A.M. so I'd have a full day, if needed, for this hunt. I arrived at my starting point in the Arnot Forest shortly after 8:30 and was delighted to find, right where I had hoped to start the hunt, some white sweet clover (*Melilotus alba*) plants bearing fresh blossoms and being worked by honey bees. Soon three worker bees were buzzing in my bee box, ready to be introduced to my comb filled with anise-scented sugar syrup. When this was done, all three bees loaded up on the rich food inside the darkened box, then flew away promptly when set free, and soon came back for more. Great! This told me that there was not a honey flow on, so I should have no difficulty getting a heavy traffic of foragers coming and going at my comb.

Fig. 5.4 shows the notes I made in my notebook between 9:55 and 10:33, when I was at the starting point for this hunt.

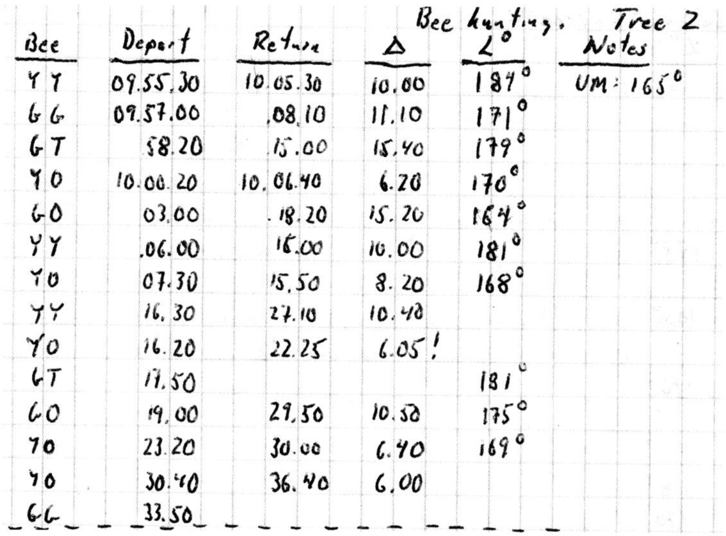

Bee	Depart	Return	Δ	Bee hunting. ∠°	Tree 2 Notes
YY	09.55.30	10.05.30	10.00	187°	UM: 165°
GG	09.57.00	.08.10	11.10	171°	
GT	58.20	15.00	15.40	179°	
YO	10.00.20	10.06.40	6.20	170°	
GO	03.00	.18.20	15.20	164°	
YY	.06.00	15.00	16.00	181°	
YO	07.30	15.50	8.20	168°	
YY	16.30	27.10	10.40		
YO	16.20	22.25	6.05!		
GT	17.50			181°	
GO	19.00	29.50	10.50	175°	
YO	23.20	30.00	6.40	169°	
YO	30.40	36.40	6.00		
GG	33.50				

FIG. 5.4. Notes taken at the start of a bee hunt on the morning of July 30, 2011. One bee, YO (yellow-orange) needed only about 6 minutes to fly home, unload, and fly back to the feeder. All the bees flew off in essentially the same direction, nearly due south.

They show that I had five bees labeled with spots of paint for individual identification and that these bees were making regular trips to my comb: YY (yellow thorax, yellow abdomen), YO (yellow thorax, orange abdomen), GT (green thorax only), GG (green thorax and green abdomen), and GO (green thorax and orange abdomen). The notes also show that all five bees flew off in about the same direction, approximately 174°, which gave me confidence that all were flying home to the same bee tree. Several unlabeled bees were also visiting the comb, and I noted the vanishing bearing of one of these unmarked (UM) bees: 165°. Evidently, this anonymous bee was flying home to the same tree as the five identifiable bees.

The times spent away from my feeder comb, however, varied markedly among my five labeled bees. Four of them—YY, GT, GG, and GO—were always gone for more than 10 minutes.

But one bee, YO, never spent more than about 6 minutes away, which indicated that she was making quick turnarounds each time she got home. She was probably dashing into the nest, quickly offloading her droplet of sugar syrup to a receiver bee, and then rushing back out to make another foraging trip. No dillydallying at home for this industrious bee! Applying the guidelines mentioned above, I estimated the distance to the bee's nest at approximately 0.5 miles. Now I was optimistic about finding the bees' home by the end of the day, and indeed, at 4:33 in the afternoon, after moving my comb five times to stands closer and closer to the bees' home (using methods I will describe in the next chapter), I found the front door to their residence when I spotted bees zipping in and out of a knothole 44 feet up on the south side of a massive red oak, perhaps the largest in the area. When I plotted this tree's location on my topographic map and then precisely measured its direction and distance from my starting point, I was delighted with the results: 178° and 0.56 miles. The bees' vanishing bearings had guided me precisely in the direction of their home, and the shortest of their away times had revealed the distance to their home.

How did I come up with my guidelines for estimating the distance to a bee tree based on the minimal away times of bees visiting a feeder? The first step was to make careful measurements of the flight speeds of bees traveling to and from a feeder filled with sugar syrup (see biology box 5). Separate measurements were made for outbound bees, flying nearly empty, and inbound bees, carrying home heavy loads of the syrup. We found that the cruising flight speeds of our bees were 20.8 and 14.6 miles per hour—that is, 0.35 and 0.24 miles per minute—on their outbound and inbound flights, respectively. We also measured how much time an arriving forager spends making maneuvering flights when she arrives at a food source and starts work there (about 10 seconds, or about 0.2 min) and when she reaches home and prepares to land at her hive's entrance (about 25 seconds, or 0.4 min).

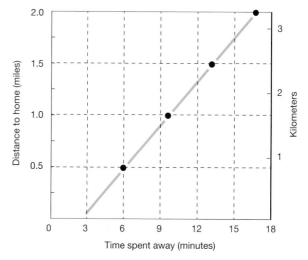

Fig. 5.5. Relationship between the time a bee spends away from the feeder and the distance to her home. This graph provides a good estimate of the distance if applied to the shortest away times observed, such as those of the bee YO in fig. 5.4.

Knowing these facts about the travel habits of foraging bees, and assuming that a forager spends only a minimal time (2 minutes) inside the nest between foraging trips, we can calculate that if the feeder comb is only a few feet from the bees' home (so the flight distance is 0.0 miles), then the minimal time that a foraging bee will be gone from your comb when making a trip home will be about 2.6 minutes (2.0 minutes spent inside the nest plus 0.6 minutes spent maneuvering at the nest and at the feeder). Likewise, we can estimate that if the feeder comb is precisely 1 mile from the bees' home, then the minimal time that a foraging bee will be gone from your comb will be about 10 minutes (4.2 minutes for the inbound flight, 2.0 minutes for the time inside the nest, 2.9 minutes for the outbound flight, plus 0.6 minutes spent maneuvering at the comb and at the nest; the total is 9.7 minutes).

Fig. 5.5 summarizes the results of these calculations in a way that makes it easy to estimate the comb-to-nest distance once you have figured out how quickly your speediest foragers are making their round trips home. This figure also shows at a glance why away times of 2–4 minutes are exciting, ones of 5–9 minutes are still highly encouraging, and those of 10–14 minutes indicate that a lengthy (but doable!) hunt lies ahead.

BEGINNER'S LUCK: THREE SHORT HUNTS

Back in the summer of 1978, when I was gaining my first experiences as a bee hunter by lining bee trees in the Arnot Forest, I had some incredible beginner's luck: three times I happened to set up my feeder comb so close to a bee tree that I soon saw bees making round trips to their home and back in less than 3 minutes. In each hunt, I found the bee tree within about an hour of introducing the first bees to my feeder comb, and I never needed to move the feeder.

One of these hunts, my shortest ever, happened on September 6, 1978. I started it in an abandoned pasture that sits on a shelf of land beside the stream that defines the boundary between the Arnot Forest and the Cliffside State Forest, in the southwest corner of the Arnot Forest, above the Opalescent Falls. At 10:18 A.M., I diligently recorded in my notebook that I had captured and released 5 bees. At 10:35, I noted that I had captured and released another 5 bees, and I included the observation "It is not hard to find bees. Is there a bee tree nearby?" At 10:40, I watched my first labeled bee fly off at 245°, straight toward the glacially steepened hillside for which the Cliffside Forest is named. (At its base, beside the stream, this hill rises 150 feet in just 80 feet, so its slope is a ladder-like 62°.) Soon I began recording the departure and return times of 7 labeled bees, and by 10:49 I was able to begin calculating the away times of these bees. Five were less than 3 minutes: 2 minutes and 30 seconds, 2 minutes and 10 seconds, 2 minutes and 5 seconds, 2 minutes flat, and 1 minute and 50 seconds. At 11:06, I jotted down the note "Left station to find bee tree," and, at

11:16, I followed up with "Maple, about 100 m [330 feet] away to the west, on the other side of creek bank!" It was a ridiculously easy find, for I was able to see the bees fly straight from my comb, over the field, across the stream, and into a knothole in an ancient sugar maple tree (*Acer saccharum*) in full view across the creek. While it is true that I spent an hour hiking to and from the remote starting point for this hunt, the period of time in which I was actually engaged in hunting the bees—from capturing the first foragers to spotting their nest's entrance—lasted a mere 58 minutes, so I figure this hunt was completed in less than an hour.

This quick success in finding a bee tree followed two similar experiences I had enjoyed just days before. On September 2, 1978, I found bees collecting nectar and pollen from goldenrod flowers (*Solidago canadensis*) near the top of Irish Hill. I got a beeline going with these bees, recorded times away from the feeder as brief as 1 minute and 45 seconds, and lined my way to a quaking aspen tree (*Populus tremuloides*) 160 feet from my starting point. Total time of hunt: 1 hour and 19 minutes. Number of moves: 0. Three days later, on September 5, 1978, I caught bees foraging on jewelweed flowers (*Impatiens capensis*) growing along the banks of Banfield Creek near the entrance to the Forest. From here, I established a beeline, recorded times away from the feeder as short as 2 minutes and 40 seconds, and traced the bees to a magnificent sugar maple tree 340 feet away. Total time for this hunt: 1 hour and 8 minutes. Number of moves: again 0.

MY DEEPEST SENSE OF ACCOMPLISHMENT AS A BEE HUNTER

The three quick triumphs as a novice were wonderful for building my confidence that I could someday become a mighty bee hunter, and I still remember vividly the thrill of discovering these trees so speedily and without assistance. My deepest sense of accomplishment as a bee hunter, though, came from a long and challenging hunt, one in which I drew on the craft and knowledge gained from long experience. I remember starting this hunt on a Sunday afternoon (August 7, 2011), when I captured bees from goldenrod plants growing on the edges of a log landing in the northeast corner of the Arnot Forest, just inside its north entrance. The beeline pointed 150°, toward the summit of Recknagel Hill, three quarters of a mile away. My next task was to collect data on the bees' departure and arrival times to determine their away times and thereby estimate the distance to their abode. Would it be on the north slope of Recknagel Hill, which rose before me, on its distant summit, or somewhere down the far side? By 1:00 P.M., I had 7 bees labeled with bright dots of pink, orange, red, and green paint, and by 1:30 I had 16 away times recorded in my notebook. They ranged from 7 to 35 minutes. The average of the four shortest ones was 8 minutes and 15 seconds, which indicated that the tree was about three-quarters of a mile away (see fig. 5.5), thus somewhere near the top of the hill.

Thunder started rumbling in from the west around 1:45 P.M., and by 2:00 it was raining hard. But knowing that when it rains hard, it rarely rains long, I built a little shelter of flat stones to protect the sugar syrup in the feeder comb (fig. 5.6) and waited in my truck for the storm to pass. By 3:05, the sun was shining again, the bees had returned to see if more loads of sweet syrup could be gathered, and soon I again had a roaring line of bees.

Now I was ready to perform the first of the seven moves of the feeder comb that I would need to make to find this bee tree. Around 3:45, I moved my feeder, plus all the labeled bees, three-tenths of a mile. I did not, however, relocate my operation to a spot straight down the beeline (direction: 150°), for doing so would have taken me directly into a deeply wooded area of the Arnot Forest. Instead, I moved off along a bearing

FIG. 5.6. Improvised shelter for the feeder comb during a thunderstorm. It provided protection to the bees and prevented dilution of the sugar syrup.

of 105°, which took me to a bluff within a 100-acre field that glowed with goldenrod flowers and provided sweeping views to the east, south, and west. This vast wildflower meadow had once been a hay field and pasture, part of a hilltop dairy farm that bordered the Arnot Forest, but in 2006 it became the Greensprings Natural Cemetery Preserve. It offers a simple, sustainable, and (I think) beautiful alternative to conventional burials. I affectionately refer to Greensprings as a "composting cemetery," and I figure that someday I will have a green burial here, with a sugar maple tree marking my grave.

This day, however, I was only passing through the place since I was hot on the trail of a wild colony of honey bees. My rationale for moving the feeder comb to the cemetery was to gain a vantage point from which I could launch a series of advances toward the top of Recknagel Hill along a route that intersected some clearings that are the remnants of long-abandoned fields. This approach would be vastly easier than moving straight up through the mature forest that covers most of the hill. The bees, obviously desperate for good forage, stayed with me during my lateral move to the cemetery, and soon I had a fresh beeline pointing 165°. I tended to the bees for about an hour at this new spot, to both check that all 7 of my labeled bees had shifted to the new feeder site and collect new readings of the bees' away times. Now they averaged only 7 minutes and 6 seconds. Progress! In early evening, I headed home, leaving behind in the meadow my bee box on its little yellow table, and with its door wide open, so the bees could visit the syrup-filled comb that I had tucked inside the box for the bees to enjoy.

Next morning, I was back in the meadow a little after 10:00 A.M. The day had started cloudy and cool, but now it was sunny, and the wildflowers in the cemetery's meadow were alive with insects, including two of my favorites: gorgeous yellow jacket wasps and eye-catching red-and-black milkweed bugs. With me was a friend, Sean Griffin, who had worked with me on various bee projects over the past four years while an undergraduate student at Cornell, and who would soon be starting graduate studies in pollinator ecology at Rutgers. Sean was helping me locate 10 bee-tree colonies in the Arnot Forest that summer for a study of the genetics of the wild honey bees living there, and already had impressed me with his talent as a bee hunter.

The feeder comb was empty and abandoned when we arrived at the bee box that I had left behind in the cemetery, but 2 bees arrived while I was refilling it. While they were loading, Sean and I hiked down the line to find a good place for our next stand, and we found an open spot, beside a small pond, about 270 yards away in the direction the bees were flying. By 11:35, the bees were mobbing the comb and we were refilling it frequently, so it was easy to catch 10 or so bees in the box by sliding the comb inside and snapping the door shut. We then gathered up our equipment, moved to the pond, and released the bees. Fortunately, the bees accepted the move, and by 12:15 we made another 270-yard jump down the line, this time landing in a clearing that we called "the meteorite" because on its edge was a rounded, alien-looking, jet-black boulder that was nearly as tall as I am. I suspect that it is a glacial erratic plucked from the Adirondack Mountains, more than 100 miles away, and left behind when the latest glacier melted away some ten thousand years ago.

At this point, we had moved about a third of a mile from our overlook site in the cemetery meadow, had dropped down in elevation by 150 feet, and had reached the edge of the deep woods. No more clearings were available. Moreover, in another 120 yards we would reach the rim of a gorgeous, 60-foot-deep chasm lined with shady hemlock trees; it had been dug by a tributary of the Jackson Creek, a bigger stream about a mile to the east. When released in the woods, bees generally circle up and clear the tops of the trees before striking off toward home, so it can be hard to tell the direction they are traveling. This means that when you get into the woods, you must look hard to find places where there is a canopy gap large enough for you to follow a departing bee long enough to see what direction she heads off before she disappears. Fortunately, we discovered such a gap on the far side of the ravine, 200 yards from "the meteorite" and more or less along our beeline's bearing of 165°. By 1:00 P.M., we had jumped with a batch of bees to this site, and thankfully it was not long until we heard the welcome hum of arriving bees. Over the next hour, we were able to get several good readings of the vanishing bearings of bees, and they averaged 163°, which told us we had stayed true to the beeline bearing and had not gone past the bee's home. We also recorded an away time of just 4 minutes flat, which suggested that the tree was only about a quarter of a mile away.

Over the remainder of the afternoon, we made three more moves to new stands, each another 100 or so yards ahead of the last. Each move took about an hour, for although the bees arrived in ever greater numbers and thus there was no more worry about losing them, I wanted to determine at each site whether the bees were still heading forward or had begun going back. Seeing the latter would tell us that we had passed the tree. I was glad Sean was along, because while I patiently kept the feeder comb stocked with sugar syrup and strained my eyes trying to track the disappearing bees long enough to see where they were heading, Sean—being younger and more impetuous—would scramble up the hillside examining every likely tree. Eventually, we finished climbing the steep side of Recknagel Hill and found ourselves not heading for its summit but traversing a gently sloping shoulder, moving along the same 165° bearing we had followed since the start of the day. I made the last move at 4:35, with dozens of bees buzzing round my head and mobbing the dropper whenever I removed it from the syrup bottle to refill the comb. The bees were mad for the anise-scented syrup, and I was convinced the tree must be close by, so I joined Sean in making a tree-by-tree search. For me, this is often the hardest part of the hunt. Nothing is more frustrating to me when bee hunting than knowing that I am close to a bee tree but haven't yet spied the telltale flash of wings in the air outside the nest entrance. So it was a profound delight when, at 4:58, Sean ended my bafflement and doubts by shouting, "Found it!"

Searching approximately 100 yards beyond the feeder, Sean had spied bees swirling outside a knothole 37 feet up on the southwest side of a towering sugar maple tree (fig. 5.7). Once you were looking in the *right place*, you might be tempted to say that the bees and the nest entrance were in plain view. But here on Recknagel Hill, with its thousands and thousands of mature hardwood trees, each one a prime candidate for containing a nest cavity attractive to honey bees, finding the one *right place* to look to see the bees was a daunting test of a bee hunter's skill. Passing this test gave both Sean and me an unforgettable thrill and a deep-seated sense of accomplishment.

Fig. 5.7. Sugar maple tree atop Recknagel Hill that houses a colony of honey bees. The nest entrance is 37 feet off the ground, which is not unusually high. Sean Griffin looks up at his find.

BIOLOGY BOX 5
How Fast Do Honey Bees Fly?

In the summer of 1983, I measured the flight speeds of bees flying to and from a feeder that provided a profuse supply of a sucrose solution, much like the comb of a bee hunter. I did so for a study of how bees adjust the strength of their waggle dancing in relation to the energetic profitability of a nectar source (Seeley 1986). It is easy to determine how many joules of energy a foraging bee *gains* when she collects a load of nectar, for one can easily measure how much sugar solution she imbibes and what its sugar concentration is, and these two variables determine the energy content of the bee's nectar load. Calculating how many joules a foraging bee *expends* when she collects a load of nectar, however, requires knowing how long she flew to collect the load and what her metabolic rate was while she was in flight. And to know how long a flying bee takes to get to and from a food source, one needs to know how fast bees fly.

To see how fast nectar foragers fly, I placed a hive of bees at the north end of a narrow hay field that stretches for nearly 1,000 meters (0.6 miles) along the top of an esker in the Yale Forest in northeastern Connecticut. All around this hay field are woods and swamps for several miles. I then trained 25 foragers from my hive to collect a concentrated sugar solution (2.5 molar sucrose) from a feeder that was located 500 meters (1,640 feet) to the south, hence partway down the field. Each bee was labeled for individual identification with paint marks on the thorax and abdomen. I sat at the feeder while an assistant sat at the hive, but we were in close communication with walkie-talkies. To measure how long it took bees to fly from the hive out to the feeder, the assistant would signal me the moment a labeled bee left the hive, at which instant I would start a stopwatch that I would let run until I saw the labeled bee arrive at the feeder, whereupon I would stop the stopwatch and record the bee's flight time out to the feeder. Then, when this bee had finished loading up at the feeder and had taken off to fly home, I would start the stopwatch and alert my assistant to look for the returning bee. When he reported that the bee had landed at the hive, I would stop the stopwatch and record the bee's flight time back to the hive. We did this for all 25 bees.

Once we had data for all the bees with the feeder at 500 meters, I gradually moved the feeder—with the bees tagging along—farther from the hive until it was 700 meters (2,256 feet) from the hive. At this point, my assistant and I repeated the flight time measurements for the 25 bees. Once we had done this, I again moved the feeder to the 900 meter (2,952 feet) mark, and we once again measured the bees' flight times. The results of these measurements are shown in the figure, in which two lines relate flight time to flight distance—one for the bees' outbound flights and the other for their inbound flights. The slope of each line indicates the average cruising flight speed of the bees: 9.5 meters/second (20.8 miles/hour) for their outbound flights and 6.7

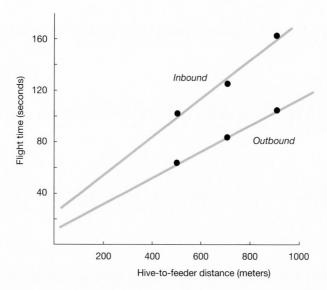

Relationship between flight distance and flight time for the inbound (homebound) and outbound flights of bees collecting large loads of sugar syrup from a feeder.

meters/second (14.6 miles/hour) for their homebound flights. The point at which the lines intersect with the vertical axis shows how much time the bees spent at the end of each flight, either maneuvering to land at the feeder (end of outbound flight, 10 seconds) or at the hive (end of homebound flight, 25 seconds).

Why was the average flight speed of the bees so much higher on the way out to the feeder than on the way home? It is not because the bees enjoyed a tail wind on the way out and fought a headwind on the way home; the data were collected on two virtually windless days. In a later study (Seeley 1994), I found that the mean weight of worker bees visiting a feeder loaded with a 2.5-molar-sucrose solution was 76 milligrams when they landed at the feeder, and then was 138 milligrams when they started to fly home from the feeder. Therefore, on average, each bee was airlifting home 62 milligrams (= 50 microliters) of the 2.5-molar-sucrose solution. In other words, these bees were flying home with payloads equal to 82% of their body weight! It is no wonder they flew home with only 70% of the flight speed they had when they flew (nearly empty) out to the feeder.

 CHAPTER 6

Making Moves Down the Beeline

At this point in the hunt, you have established your beeline, so you know the direction to your bees' home. You have also estimated the distance to the bees' home, having timed several bees, seen how long they were away from your comb, and then translated this time into a distance. So now you are ready to start making moves down the line. You might wonder why, knowing the direction and distance to your bees' residence, you don't proceed directly to the area indicated and start hunting for the bees. The reason is simple. Unless you have been fantastically lucky and begun your hunt only a hundred or so yards from the bee tree, you won't know the direction and distance to your bees' abode with any exactitude. In most cases, your sightings and timings will leave you with a mystery whose solution sits somewhere within an area about a quarter-mile square. If this area is forested, then it will contain thousands of trees, far too many for you to have a decent chance of discovering by direct search the one occupied by your bees. As we saw in the last chapter, even when you have narrowed your search to a comparatively small area of a hundred yards square, which contains only a hundred or so large trees, it can still be wickedly hard to spy the knothole or crack, usually high in a tree, that is the entrance of your bees' dwelling place. You do your-

self a huge favor by making moves down the beeline, thus letting the bees guide you, step-by-step, back to their otherwise
secret residence.

To make a move down the beeline, you first place the bee
box on your stand and slide a syrup-filled comb into the front
compartment with the door left open. (You should hide your
other comb; I like to cache it in a Ziploc storage bag.) Great
confusion on the part of the bees ensues. They circle your box,
looking for a comb sitting as plain as plain can be on your
stand. As more bees arrive, the air fills with suspicious bees reluctant to enter your box. Sooner or later, though, the tempta-
tion of another swig of your sugar syrup becomes too great and
one bee alights on the comb. Soon others drop down, probably
emboldened by the first bee (fig. 6.1). When a half dozen or
more have settled and begun loading, gently close the door to
trap them. Next, lure your imprisoned bees to the window in
the rear compartment and then lower the sliding divider to
confine your captives in the back chamber. Once they are
locked up in the rear compartment, reopen the door and capture another bunch of bees in the front compartment. Gather
up as many of your *labeled* bees as you can so you don't lose
them during the upcoming move. By now, you've probably
come to know the various work habits of these individuals—
some are brisk and efficient, while others operate in low gear—
and you certainly want to keep your star performers engaged
in your hunt. The final preparation for the move is to put a
rubber band around your bee box so the front door, sliding
divider, and window shutter are all secured for the move.

Now gather up your paraphernalia and hike 100 to 300
yards down the beeline to a new clearing, one that you have
scouted out in advance of the move. Once there, set up your
stand and release the bees from the front and rear compartments. Releasing the two groups separately will give you two
opportunities to watch your bees fly home.

Soon all the bees have flown off, leaving you all alone at
your little filling station. You are still open for business, but of

FIG. 6.1. The comb has been placed in the outer compartment of the bee box, and bees have settled on it to fill up on the sugar syrup. When the door is closed, the whole group will be captured for the move to a new site down the beeline.

course, having relocated, you've lost some customers. Without a steady traffic of bees stopping in for quick refills, the place is quiet. Too quiet. You begin to wonder, will any of the bees come to this new spot, or will they simply go back to your original location? That is, after all, where they experienced fine foraging for the last hour or two. You check your watch . . . maybe five minutes have passed . . . still no bees.

This is one of the most suspenseful, moments in a bee hunt. If you have mistaken the bees' flight route home, and so have drifted too far off the beeline to the right or to the left, then the bees might not return. This can also happen if a honey flow has begun. In this case, your colony's food-storer bees will be so busy handling the flood of nectar brought home by your colony's other foragers that your bees will experience long delays in unloading their cargoes. If no bees come back within 15 minutes, then you will need to hike back to your starting location, box up another batch of bees, and try again to make the move. But let's assume that all is well—you have moved squarely down the line and the flowers are not offering stiff competition—and soon you hear the silvery tone of a returning bee, followed shortly by another and another and another. Fabulous! The bees have made the move.

MAKE MOVES UNTIL THE BEE TREE IS REACHED OR PASSED

Now success seems assured. In principle, all you have to do is to continue making moves down the line until the tree is reached or passed. If you overshoot the bees' home, then your beeline will reverse direction, telling you that the bee tree is somewhere between your present and previous stands. In practice, however, your skill and perseverance as a bee hunter can be tested severely when you execute a series of moves down the line. It is these challenges that make bee hunting the art and sport that it is.

To see how the move-down-the-line phase of a bee hunt can test one's mettle, let's review some statistics on making moves that I collected when I made a census of the wild colonies of honey bees living in the Arnot Forest (see Seeley 2003a, 2007). In 2002, between August 20 and October 1, I hunted in this forest for 117 hours spread over 27 days, during which time I established lines from 12 clearings spread over about half of the forest (fig. 6.2). At four of the sites (4, 5, 6, and 9), the beelines pointed either toward bee trees that I had already located or ones that I judged easier to approach from a different location, so I did not pursue any of the lines established at these sites. At all of the other eight sites, however, I established at least one strong line to a colony of bees living less than a mile away, and from each of these places I launched a full-blown hunt. All eight were successful.

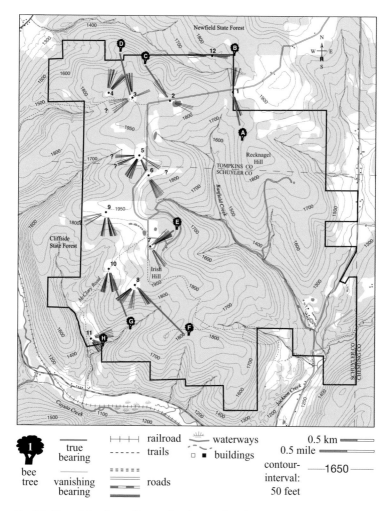

Fig. 6.2. Map of the Arnot Forest, showing the locations where beelines were established (1–12) and where bee trees were discovered (A–H). The base of each bee-tree symbol marks the location of the bee tree.

Because I was conducting these hunts for a scientific study, I took detailed notes on the work, and from these notes I have calculated the statistics presented in table 6.1. On average, I worked some four-tenths of a mile down any given line to reach the home of the bees I was trailing. Also, on average, I covered this distance by making 4.6 moves, each one some 60–400 yards long, and I spent about an hour at each location I moved to—that is, each new stand. Some of the time at each new stand was spent waiting for the bees to find their way back to the site, but most of it was devoted to collecting data on the bees' vanishing bearings and away times. Once I had determined the direction the bees were disappearing and had updated my estimate of the distance to their home, I would spend time searching in the indicated direction. Usually, I would be trying to find a good clearing for the next stand. If, however, I recorded away times of just 2–3 minutes, then I would switch to inspecting every likely tree for the bees' dwelling place. The range of my searches was either 100 or so yards ahead or behind, depending on whether the bees were flying off in the same direction I had been moving or were circling back toward where I had come from. The latter behavior signaled that I had moved too far. On three of the eight hunts, I did make a move beyond the bee tree and so saw the beeline reverse itself. Finally, as shown in table 6.1, I typically searched for about 2 hours (113 minutes, on average), to discover the bees' nest site after I had determined that it must be nearby and began making a tree-to-tree search for my prize. How one conducts the final search is the subject of the next chapter.

Table 6.1 also shows that on average I spent more than 10 hours, spread over 2.4 days, to find a bee tree. This makes an important point: bee hunting takes time. You must be prepared to spend more than one day to find a tree. That said, I should explain that the values of 10.4 hours and 2.4 days are somewhat misleading because they are inflated by the circumstances in which I conducted these hunts. First of all, I was hunting in the fall, when the nights were chilly, so on many days the bees started appearing on the flowers or back at my feeder comb only around mid-morning. It was impossible to make a good, early start. Second, the difficulty of making only late starts was

TABLE 6.1 Mean Distance and Time Variables for Eight Successful Bee Hunts in 2002 in the Arnot Forest

Total distance of hunt: 0.41 miles (range 0.16–0.70 miles)

Time spent per hunt: 10.4 hours (range 7.5–10.7 hours)

Days per hunt: 2.4 days (range 1–6 days)

Moves per hunt: 4.6 moves (range, 3–11 moves)

Move length: 127 yards (range 60–400 yards)

Time spent at each moved-to site: 57 minutes (range 25–92 minutes)

Time spent searching for tree from final site: 113 minutes (12–170 minutes)

compounded by the fact that the classes I was teaching at Cornell started around midday. So on weekdays, I was free to go hunting for only a few hours in the afternoon. (Sometimes, however, I gave myself the pleasure of popping out to the Arnot Forest early in the morning to reload the comb, to sustain the bees' interest in my feeding station until I could get to it around 2:00 P.M.) If I had not had these midday commitments, then I'm sure that I would have needed fewer than 2.4 days, on average, to complete these hunts. I often get my bee tree in just 1 day when I have no disruptions.

Making the moves down a beeline can test not only your patience and perseverance, but also your skill and cunning. For one thing, in making a move down the line, you should set up the next stand in a clearing, but sometimes there is no open area dead ahead. So you explore more widely for a good place. You might find a nice open place a bit off the line, and as long as it is not far off, the bees will accept moving to it, especially if they are desperate for food. You might find, however, that you have reached a place where the forest cover is solid. If so, then you face a challenging situation because unless the bees are within about 100 yards of their home, they will travel to and from your filling station by flying above, not through, the trees. This means that when you release bees in a spot without a gap in the canopy, the bees will circle up and disappear in the tops of the trees, making it impossible to tell whether they are flying forward or backward to get home. In this situation, you use whatever gap in the canopy you can find, and you do your best to determine on which side of this opening the bees leave the area. The most important thing is to move as straight as possible down the line. Edgell (1949) stated this clearly: "If you meet a swamp, you must go through it. If you meet a cliff, you must go up it. If you meet a pond, you must go round it and set up at just the right point on the opposite side." I agree.

WHEN THE BEELINE REVERSES ITSELF

One of the most auspicious experiences you can have during the move-down-the-line phase of a bee hunt is to see your beeline reverse itself. This usually happens when the bees are foraging steadily and the moves are being made easily. If you are making short moves, then some of the bees might even follow you to the new stand so that shortly after you release the captured bees from your box you will soon see others, lured along by the scent, settle on the comb and begin to load. Now you have no worries about losing the bees, and you can make your moves quickly. You can also begin to hope that the bee tree is in sight. At some point, you might make a longer move, hoping to speed things up, and you find that now your labeled bees are a long time in coming back. You might also notice that when they do eventually reappear and finish loading, they behave oddly. Instead of flying straight away in the direction you have been moving, they circle off in all directions. After making a few round trips, some of them begin to fly straight away, but they are heading back in the direction of your last stand. Your beeline has reversed itself!

There can be no doubt, the bees' home is somewhere between your current stand and the previous one. Usually, the bee hunter now has only to carefully inspect every tree back down the line to discover the bee's home. Sometimes, if the nest's entrance is in plain view on the side of a tree's trunk, you can spy it simply by looking back in the direction you just came from. I had this experience once when lining bees in the Arnot Forest in 2002. On Tuesday, September 24, early in the afternoon, I had established a beeline with bees caught off goldenrod plants that I found in a clearing where once had stood a farmhouse. This site was in the southwest corner of the Arnot Forest, just 200 yards from where the abandoned McClary road exits this forest (site 11 in fig. 6.2). The beeline pointed 139°, and the minimum away times were just over 3 minutes, so I knew the bee tree was only about 0.15 miles (800 yards) away. Most promising! Over the next 2 hours, I made 2 moves totaling 600 yards along a bearing of 140°. This took me out of the old farm field and into the woods at the top of a 60-foot-deep ravine carved out by a stream that flows from northeast to southwest, hence perpendicular to my beeline. From my stand at the top of the north side of the ravine, I could watch the bees slowly fly off along a bearing of 129° (straight across the ravine), and I recorded two away times of just 2 minutes 15 seconds and 2 minutes 45 seconds. I knew, therefore, that the bees' nest was nearby, so I worked my way down the side of the ravine, inspecting carefully every tree along the line of the bees' flights, but I could not find their home before dark. This left me thinking that the bees might be living along the top of ravine on its far side, some 150 yards away as a bee flies.

I returned to the stand early the next morning, refilled the comb with sugar syrup, quickly reactivated a team of foragers, and then moved my operation to the top edge of the ravine on its south side. Now what? Well, now the bees flew away with an average direction of 315°, hence 186° from the previous 129° bearing. The line had reversed! At this point I was certain that the bees were living somewhere down in the ravine, but now I also had to dash back to Cornell to teach my animal behavior class. So although it pained me mightily to do so, I broke off my hunt. But by 2:30 p.m. I was back in the forest and climbing from my car parked by the gate at the top of the McClary Road. By 3:02, I had dashed the mile and a half to the ravine and had scrambled down its north side and up its south side. It was a sunny day, and the cherry, maple, and oak leaves had started to fall, so the woods were brightly lit. Things were looking good. By 3:30, with the comb reloaded, I again had a heavy traffic of bees, and at 3:50 I moved my operation two-thirds of the way down the south side of the ravine, along a 320° bearing. Perched there, I had a perfect view of the departing bees' flight paths, which shot out across the ravine. They angled upward only slightly, hence neither up to the treetops nor down to the streambed. Soon I spotted the nest entrance. It was in clear view, 14 feet up on the southwest side of a white ash tree (*Fraxinus americana*) that stood 40 yards away, on the north side of the ravine, and in the direction of 319° (fig. 6.3). As always, I was thrilled by my find, but this time I was not entirely proud of my accomplishment. Now that I had found the bee tree, I could plainly see that when I had descended the north side of the ravine that morning and that afternoon, making my way down along a deer trail, I had actually walked within 10 feet of the tree without noticing the bees. How embarrassing. Good thing I was hunting by myself that afternoon.

I will share a second example of how I've made good use of a reversed beeline, this time in locating a most peculiar bee tree. In August 1981, I was living in New Haven, Connecticut, and had finished my first intensive year as an assistant professor in the Biology Department at Yale University. I needed to get out of the city and into nature to gather my wits before the start of the new academic year. My wife, Robin, a marine ecologist who was at that time working on her PhD in biology at Yale, suggested that I go bee hunting in the Yale Myers Forest, a

Fig. 6.3. White ash bee tree, found in the Arnot Forest on September 25, 2002. The height of the man standing beside the tree, Dr. Jun Nakamura, is 5 feet and 7 inches. Brown propolis stains mark the entrance openings of the bees' nest.

7,840-acre forest preserve surrounded by vast state and private forests in Tolland and Windham Counties in northeastern Connecticut. This would give me a perfect opportunity to start exploring this forest as a possible location for field experiments with the bees that I was planning for the following summer. Dr. David M. Smith, professor of silviculture in Yale's School of Forestry and Environmental Studies, encouraged me to work in the Forest, said it was OK to stay in the bunkhouse, and gave me a map of the place. The next day I headed north, equipped with a sleeping bag, a box of groceries, and my bee box and other bee-hunting gear.

On my first afternoon at the forest, I searched for honey bees foraging in the white clover growing in the lawn in front of the bunkhouse. My goal was to get a beeline going to start mapping out the wild colonies living in the area. I wanted to get a feel for the abundance of the honey bee colonies living in this forest, to learn how much interference I might get from their foragers when I would conduct behavioral experiments there (see biology boxes in chapters 1 and 5). I quickly spotted a worker bee, and soon I had her trapped in my bee box and loading up on my anise-scented sugar syrup. When released, she circled the box, memorizing its location, and then flew slowly off to the east. Ten minutes later she was back. Within the hour, I had 7 bees labeled with paint marks visiting my comb. Their minimum away times were 6 minutes, which told me that their home was only about a half mile away. Over that afternoon and the next morning, I worked my way steadily toward the bees' nesting site in a series of moves, one of which was a spectacular 300-yard jump across a swamp. This took me to a densely wooded area at the foot of Vinton Hill. Normally, coming to unbroken woods would slow me down and require short moves, but in July that year gypsy moths (*Lymantria dispar*) had defoliated the forest canopy in northeastern Connecticut, and in August the woods were so leafless that they looked as they had back in April, before leaf out. So it was easy to track the departing bees through the trees, and I quickly zeroed in on the patch of woods where the bees were living.

But there was something odd going on here. I was seeing bees flying away from my comb along a bearing of approximately 100°, and many of these bees were real speedsters, making round trips to home and back in under 2 minutes and 30 seconds. This told me that their nest was toward the east and must be within sight, but I saw no large trees nearby in this direction. I searched everywhere but could find no traffic of bees diving into an opening, either in a tree or in the ground. Where in this patch of young forest could these bees have built their home? I was mystified, so I decided to check my estimation that the bees were living close by. I did so by making another move and seeing if the beeline would reverse. When I moved to a new spot just 80 yards ahead, sure enough, the bees reversed their flight direction, heading off at 280°. What on earth was going on here? Curiously, the bees' flight line now pointed straight at the only tree in this patch of woods that had escaped defoliation by the gypsy moths: a thick-crowned eastern hemlock (*Tsuga canadensis*). But it was a rather small tree, only 13 inches in diameter at breast height and no more than 35 feet tall, thus much too small to be a bee tree. Then I suddenly realized a possible solution to the mystery: sometimes a honey bee colony will make the fatal mistake of building its nest out in the open rather than inside a protective cavity (fig. 6.4). Almost always, where I live, an open-nesting colony will perish over the winter from wind exposure. Had a swarm of bees taken up residence in the dense, dark crown of this hemlock tree, attaching their combs to the tree's branches? I quickly walked over to the tree and examined its upper branches from below rather than from the side, and, yes indeed, I could see a mass of beeswax combs the size of a soccer ball, all covered with bees, about 20 feet up in the tree. These bees had hidden themselves away in an unexpected place, and they had had me

Fig. 6.4. The nest of a honey bee colony that failed to occupy a protective nest cavity.

stumped for nearly two hours, but in the end, I had guessed their trick. Checkmate!

A FEW WORDS ABOUT CROSS LINING

Starting with Mr. P. Dudley, who described in 1720 a "Mathematical way of finding Honey in the Woods," many authors of articles about bee hunting have promoted cross lining—also known as triangulating—as a smart way to find a bee tree. The gist of their method is as follows. First, the bee hunter catches several bees, baits them all with sugar water or diluted honey, releases just one bee and determines the course of her flight back to her home, and then notes her direction using a pocket compass. Let's say it is north (flight direction = 0°). Second, in order to determine the exact distance to the bees' home along the first bee's flight line, the bee hunter makes an offset of a quarter mile or so at a right angle to the first bee's flight line, then he lets another bee go and observes her flight course care-

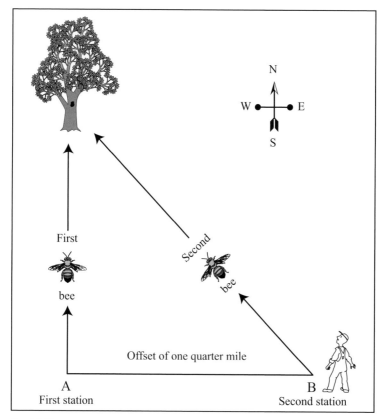

FIG. 6.5. The process of locating a bee tree by cross lining, also known as triangulating.

fully as she too flies directly to her home. Let's say the bee hunter moved east by a quarter of a mile and saw his second bee depart to the northwest (flight direction = 315°). At this point, there remains nothing to do but determine where the two lines intersect, which is one quarter of a mile north of where the bees were caught (fig. 6.5). There will be the bee tree. It is that simple!

The truth is, however, that there is only a tiny probability of getting precise information about the direction to any bee's abode the first time she leaves a bait station. The likelihood of getting this information is even tinier for the second bee, which has been greatly displaced from where she had been foraging and hence is almost certain to be lost when she flies off. What is absolutely certain is that she will circle widely, in an attempt to get her bearings, before she disappears. So even if she is not lost, she is not going to show you what you need to know to triangulate your way to her forest home.

Cross lining should not, however, be completely dismissed. Sometimes you will establish a line to a bee tree too far away to be worth hunting. If you try again somewhere in the general direction indicated by your first line, then you might get a second line that crosses your first line in a location not far away. If you then make a move to a clearing close to where the two lines meet, there is a good chance you will be near a bee tree. And not just in theory. The two examples given above of the use of a reversed beeline show how valuable a cross line can be to the bee hunter who knows he is close to his quarry but is having difficulty getting it squarely in his sights.

BIOLOGY BOX 6
How Do Honey Bees Find Their Way?

One of the many marvelous abilities of honey bees is their skill at finding their way between their nest and a flower patch that may be several miles away. To understand how they do so, it is useful to distinguish two parts of their navigation challenge: (1) the large-scale problem of steering a proper course out to the distant flowers (or back home to the nest), and (2) the small-scale problem, when within view of the goal, of pinpointing the location of the food source (or nest site). We now know that a worker honey bee solves the first problem using methods analogous to those used for ages by sailors making a passage over open water to reach a distant harbor: setting the course of her flight with the aid of a compass—for the bees, this is the sun—and keeping a running tally of the distance traveled. By this process, a bee maintains a continuously updated knowledge of her position relative to her starting point, which is either her nest (when flying out to a food source) or her food source (when flying back home). We humans call this method of orientation "dead reckoning." It is a means whereby a human or animal can keep track of its spatial position without relying on a map—that is, a two-dimensional representation of the spatial relationships among earth-based features of the environment. Before celestial navigation and GPS systems were developed, dead reckoning was how a sailor navigated when out of sight of land, when he could not determine his location on a map. Dead reckoning is how honey bees still solve the large-scale orientation problem, since there is little or no evidence that they are able to develop navigational maps (Wehner and Menzel 1990; Wray et al. 2008).

To solve the small-scale orientation problem, we now know that a worker bee uses methods similar to those used by a sailor when he comes within sight of his destination. First, she guides her initial approach by orienting to landmarks (for example, rows of trees) associated with her destination, just as a sailor refers to landmarks along a coastline to locate his harbor. And then, just as a sailor locates his particular dock within the harbor by recognizing its visual appearance, the bee pinpoints her final destination (food source or nest entrance) by remembering its look. To do these things, the bee refers to the sequence of visual images she previously experienced (and memorized) when she approached a flower patch or her home. As she approaches one of these goals, she compares what she currently sees with an image previously memorized, and she flies along the path that produces the best match between current and remembered images.

This process of bees finding their way by orienting to landmarks has been examined most closely for the final stage of a bee's journey, when she must pinpoint the location of her goal, such as a feeder

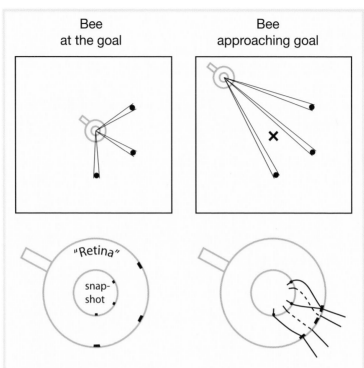

Schematic representation of how a worker bee uses visual information about landmarks to orient back to a particular goal. *Left:* when the bee is at the goal, what she sees (her "retinal image") matches what she saw and memorized (to form her "snapshot") when she was at the goal previously. *Right:* when the bee is approaching the goal, her retinal image does not match her snapshot and she adjusts her flight path to lessen the discrepancy between the two.

(Cartwright and Collett 1982). Experiments have revealed that in order for a worker bee to return to a specific spot, all she learns is the appearance of the landmarks viewed from that spot. It is as though she takes and stores a panoramic snapshot of the landmarks from that location. Then, to find her way back to that particular location, she compares the image of the landmarks that she sees (on her retina) with the image of these landmarks that she has stored (as a snapshot), and moves in a way that makes the two images match (see fig.). This is a remarkably simple mechanism for a bee to find her way back to a particular spot, for the bee does not need a map-like representation of her immediate surroundings to use nearby landmarks to locate a food source or nest site. Indeed, a bee does not even need to identify landmarks as such. The bee hunter, however, pays a price for the bee's simplification of her visual world: any alteration of the landmarks around the feeder comb—such as moving the comb inside the bee box—generates some confusion for the bee.

CHAPTER 7

Finding the Bee Tree

As you move closer and closer to the mysterious dwelling place of your bees, they will visit your feeder in ever-greater numbers, so that eventually you will have dozens, if not hundreds, of bees buzzing around you and mobbing your comb. You will find yourself refilling the comb often. You might also find yourself feeling convinced that the bee tree is close at hand, maybe even in sight! Alas, usually it isn't. A strong colony of honey bees can easily dispatch dozens of foragers to a syrup-filled comb when it is still several hundred yards away, so having a large company of bees at your comb is not a sure sign that their home is close by. What is, however, a tried-and-true indicator that your bees' abode is within a stone's throw of your comb is seeing bees that leave your comb laden with syrup and then return for more in just 2–3 minutes. If you see any of your identifiable bees commuting this quickly, you can be dead sure that the prize is somewhere within 100 yards or so of your stand. And, of course, if you make a short move and the beeline reverses, then you know with certainty that you have the tree pinned between your current and previous stands. Success is now only a matter of going down the line, inspecting every likely tree, and doing so carefully enough to discover the front door of your bees' home.

A note of caution is in order here. Once you have determined that you are indeed close to the bee tree, you should not

Fig. 7.1. Conspicuous nest entrances. *Left:* Entrance opening marked by brown tree resins (propolis) that the bees have collected and applied around the aperture to smooth out their "landing field." *Right:* Entrance opening marked by worker bees clustered outside it on a hot day.

think that the game is almost over. It might be, if the entrance opening of your bees' nest is located in plain view along the tree's trunk or an unshaded limb. Indeed, sometimes the entryway to a wild colony's nest is truly eye-catching, as when there is a ring of dark propolis (tree resins collected by the bees) around the opening or there is a throng of glittering bees clustered outside (fig. 7.1). Often, though, the game is far from over. As mentioned in the previous chapter, I usually have to search for an hour or more, going from tree to tree, to find the bees' home once I have located its neighborhood (see table 6.1). The reason this final stage of the hunt takes this long is that many nest entrances are hard to find. Indeed, some are extraordinarily hard to find. If the entrance hole is high in a tree or hidden by leaves, then your first sign of its location is likely to be a mere glimpse of sunlight flashing off the bees' wings or of bees hovering beside a tree. You must then maneuver about to find a place where you can see the bees more clearly. Did you really spy a traffic of honey bees, or was it just a few sex-starved flies chasing one other? Fortunately, most times the glitter of wings in the air will be your bees.

One scenario in which it is especially hard to bring a bee hunt to a satisfying close is when you see your homebound bees fly up into the crown of a tree, but then you cannot spot the knothole or crevice that is the entryway to their breezy residence. This leaves you with doubts that you have indeed found the bee tree. I recall, with some pride, how I once managed to overcome this particular challenge. On September 20 and 21, 2002, I worked a strong line that ran for 0.54 miles down a steep slope in the Arnot Forest, and eventually I zeroed in on a tree in which I thought the bees were nesting (see tree G in fig. 6.2). It was a strikingly tall sugar maple—I call it the Mystery Maple—growing on the far side of a stream opposite my final stand (fig. 7.2). I suspected that this tree was the home address of these bees because I had seen them leave my feeder and fly upward across the stream straight toward the crown of this tree. But I hadn't seen them disappear inside an opening in the tree. Unless I could see the bees entering some hole or fissure, I could not be certain that this tree was their home. In agony from the ambiguity, I hiked out of the Arnot Forest, made the 52-mile drive to my home and back to get a powerful pair of binoculars, and then hiked back to the tree. To my exasperation, I found that even with the help of these binoculars I could not catch sight of the bees diving into an opening anywhere within the superstructure of this tree.

The next day, September 22, I had another go at solving the vexing puzzle of where these bees were entering their nest. I had decided to make use of two trees growing near the Mystery Maple: a large hemlock growing about 20 yards from the Mystery Maple and a pole-size sugar maple growing beside the

Fɪɢ. 7.2. The Mystery Maple bee tree.

hemlock tree. By shinnying up the sugar maple, I was able to grab hold of the lowest branches of the hemlock tree and haul myself up into it. Once I was safely perched in the hemlock, I could climb its branches, almost like rungs on a ladder, to the height where I had seen the bees vanish in the Mystery Maple's crown. Looking out from my aerie, I finally spied the solution to the mystery: some 50 feet up in the tree, hidden in the crotch where the tree divided into two boles, was a crevice into which my bees were dropping to enter their home.

I have received a number of thrills in my 63 years, such as the notification in December 1977 that I had been elected to the Society of Fellows at Harvard, the citation of a gold medal from Apimondia in 1998 for my book *The Wisdom of the Hive*, and the reception in April 2001 of a Distinguished Senior Scientist Award from the Alexander von Humboldt Foundation in Germany, but, in all honesty, these thrills are all in a second class compared to the one I got when I discovered the secret passageway into the home of the bees living sky-high in the Mystery Maple.

GUIDELINES FOR THE FINAL SEARCH

The most important guideline for conducting the tree-by-tree search at the end of a bee hunt is a simple one: look everywhere! And while looking with your eyes, keep your ears open, too. Sometimes, it is possible to hear the bees buzzing, especially if it is a strong colony and it is a hot day; many bees will be fanning their wings at the entrance opening to expel hot air from the nest. The fact is, honey bee colonies take up residence in all sorts of places. Usually it is a cavity within a tree, a building, or (more rarely) a hillside or cliff face, but occasionally one finds a colony that has foolishly chosen to nest in the open air, as we have seen already (fig. 6.4). I have found nest entrances hidden in the foliage 50 feet up in the air and down in the soil between tree roots (fig. 7.3). I have found bees nesting in mas-

entrance ⟶

FIG. 7.3. Hillside bee tree in which the nest entrance is a small passageway among the roots.

sive old hemlocks, beeches, maples, and oaks, but also in slender young aspens that I mistakenly ignored at first as too small to hold a colony. I have generally found them in sturdy live trees that will provide secure housing for decades, but also occasionally in dead trees so weakened by rot that they were blown over a year or two later, and once in a hollow log lying on the ground.

In his book *The Bee Hunter*, George H. Edgell also stresses the importance of looking everywhere when you are conducting a tree-by-tree search in the last stage of a bee hunt. To emphasize this valuable piece of guidance, he quotes George Smith, the old Adirondacker who taught him the sport of bee hunting: "You look high in the maples and low in the cedars and up and down all trunks and branches, hardwood and soft, big enough to hold a hive and you can be sure of just one thing. When you do find them, they'll be where you don't expect them." A sound observation.

I should quickly add, however, that we now know that when a swarm of honey bees—that is, a group of approximately 10,000 workers bees and one queen bee that has left an established colony to start a new colony—chooses its future home-site, it does not occupy just any available tree cavity. Instead, a honey bee swarm carefully selects a nesting cavity that provides sufficiently protective and roomy living quarters. It makes this choice by means of a sophisticated process of group decision making that includes collective fact finding, open sharing of information, vigorous debating, and fair voting by the 300–500 bees in a swarm that function as its nest-site scouts. As explained in biology box 7, these scout bees prefer a tree cavity with an entrance opening that is *high off the ground* (to help avoid detection by predators, such as bears), is *rather small* (so the nest cavity is not too drafty and the nest is easily defended against small intruders, such as honey-thieving wasps), and *faces south* (so their entryway is warmed by the sun). It is not surprising, therefore, that the nest entrances of the bee trees I

Fig. 7.4. Knothole entrance of the nest in the bee tree shown in fig. 1.1, showing bees inside.

have found by bee lining in the Arnot Forest are quite high (mean height = 31 feet, range 14–57 feet), rather small (mean area = 5 square inches, range 2–12 square inches), and disproportionately located on the southern side of the bee tree (56% face SE, S, or SW, more than the 37.5% expected by chance). There is no evidence that the bees prefer entrances that are round like a knothole vs. narrow like a crack, but in the Arnot Forest, 74% of the entrance openings are knotholes (fig. 7.4) and 26% are fissures in trees.

Knowing what honey bees seek in a dwelling place is useful when you are searching for the nest entrance of a colony living in the wild. In particular, knowing that the bees are likely to be nesting high in the trees, you will want to look carefully in these high places even though they are often hard to inspect. Also, knowing that the entrance openings of the bees' homes are generally rather small—a 4 inch diameter knothole is about as big as they get—you can give just quick looks at the large openings that you find. And knowing that the bees prefer a south-facing nest entrance, you will do well to examine the sunny, south-facing sides of trees first. But do keep in mind that none of the bees' housing preferences is absolute. So when you conduct your tree-by-tree search for the entry to your bees' home, you should look high and low, south and north (and all directions in between), and for small and large. In short, look everywhere!

HOW LONG IS THE FINAL SEARCH?

One question often asked is how long it takes to find a bee tree. In my experience, it is somewhere between 58 minutes and 3 years. I have already described in chapter 5 the three instances of beginner's luck in which I started a bee hunt within sight of a bee tree and found it about an hour later. In each case, it was easy to find bees on flowers, the beeline was established quickly, there was no need to move the feeder, and when I searched down the line from the bee box, I discovered the bees' residence after inspecting no more than a dozen trees. The final search lasted less than 10 minutes in all three of these charmed hunts.

At the other end of the timescale is a bee hunt that I started in July 2011 and completed in August 2014. It began on July 31, 2011, at 11:45 A.M., when I established a beeline with two bees captured from chicory (*Chicorium intybus*) flowers beside the bridge over Banfield Creek, near the south entrance to the Arnot Forest. The line started quickly, and within 45 minutes I had given several bees paint dots for individual identification and had measured their away times. The shortest ones were 3 minutes and 30 seconds, indicating the tree was only a few hundred yards away (see fig. 5.5). Great! "This might be a short hunt," I told myself. In fact, it would turn out to be my longest-lasting search ever.

The beeline ran to the southeast, and at 1:00 P.M. I had made a 200-yard move southeast down the mostly dry bed of Banfield Creek to the point where it joins the larger Jackson Creek. From here, the bees needed just over 2 minutes to make a trip from my comb to their nest and back. Also, within 10 minutes of moving to the stony bar where the two creeks merge, I had dozens of worker bees landing at my little comb, and I was busy keeping it stocked with syrup. All this made it crystal clear that the nest of these bees was nearby, probably no more than 100 yards away.

Equally clear was its direction: 135°. This compass heading pointed squarely toward the 60-foot-high, northwest-facing slope that rises up steeply from the far side of Jackson Creek. This slope is cool and moist, and is thickly covered with soaring white pine and hemlock trees whose boughs formed a dark, bosky background against which the flying bees shone brightly in the afternoon sunlight. Soon I saw several of my bees, stuffed with sugar syrup, fly slowly and directly up the forested slope and then disappear mysteriously in the tops of some trees near its summit. I realized then that finding this bee tree would be difficult; soon I would learn that it would prove not just difficult, but almost impossible. For the rest of the afternoon I hunted for the bees' forest home by moving horizontally back and forth across the line, covering the whole area from creek to hilltop and beyond, inspecting each tree I found. I was dismayed to discover that the timber there is breathtakingly tall, and I soon found myself struggling to peer up the trunks of the towering pines and hemlocks on the moist hillside, as well as those of the maples, pines, cherries, and oaks growing on the drier hilltop. If the bees were nesting high in any of these soaring trees, then the likelihood of my spotting the entrance to their home was practically zero. I searched for 3 hours, could not find the tree, and finally gave it up.

It irked me to get so close to a bee tree but then fail to find it. So, three years later, when two of my beekeeper friends, Megan Denver and Jorik Phillips, visited me on Saturday, August 16, 2014, to show me their drone for photographing nest entrances (fig. 7.5) and to see how I hunt wild colonies of honey bees, I suggested that we make a team effort at finding the mysterious bee tree somewhere on the steep slope beside Jackson Creek in the Arnot Forest. They agreed enthusiastically, so we headed there. Starting with bees caught off Queen Anne's lace (*Daucus carota*) beside the bridge over Banfield Creek, we struck the same line as I had in July 2011. We could not tell whether the colony had survived here continuously for the past three years, or had died out and the site had been reoccupied, but either way, we knew we had the opportunity to search again for this alluring bee tree. Recognizing the difficulty of closing in on it by moving down the beeline pointing southeast, crossing Jackson Creek, and then climbing the nearly perpendicular hillside that borders the creek, we shifted the starting point of our hunt by a quarter mile to the northeast, to an abandoned gravel pit along Decker Road, a shady dirt road that climbs up Barnes Hill to the east of Jackson Creek. We hoped to have better luck in finding the bees' home by approaching it from the hillside above rather than the creek bottom below.

In the gravel pit, we caught more bees off coltsfoot (*Tussilago farfara*) flowers, and soon we had a line pointing 235°, back toward the stand of frustratingly tall trees. We also soon determined that our bees' shortest away times were just slightly over 3 minutes long, which confirmed that the nest was at most a couple hundred yards away. At this point, we started inspecting the trees down the line, one by one, but again without success. We stopped searching at the end of the afternoon, for

FIG. 7.5. Using a drone equipped with a GoPro camera to photograph a (typically) high nest entrance. Fig. 7.1, right, shows the entrance that was photographed using this equipment.

Megan and Jorik had a four-hour drive back to their home in Kingston, New York, down along the Hudson River.

That we had again come tantalizingly close to this bee tree but again had failed to find it really got my dander up, and I decided to go back and try yet again. I figured that with a touch more skill, I could shrink the search area to a small patch of the towering trees and thereby boost my chances of discovering the bees' hideaway. It rained on Sunday, August 17, but by noon on Monday it was sunny, and I motored out to the Arnot Forest to pick up where Megan, Jorik, and I had left off.

I arrived at the gravel pit a few minutes before 2:00 P.M., set out a comb loaded with anise-scented sugar syrup, and within 5 minutes one of the bees we had labeled during Saturday's hunt, Orange Abdomen, landed on the comb and began loading up on my irresistible sugar syrup. By 2:30, I had a good crowd of diligent foragers on my comb, and I was ready to start executing my plan of making a series of moves south down the dirt road. You see, when the bees flew home from the gravel pit, they did not fly down a beeline running perpendicular to the road, but I wanted them to do so now because a line running at a right angle from the road to the bee tree would have the shortest, most direct path through the woods to the bee tree. So over the next two and a half hours I made a series of four moves totaling about 500 yards down the road. When at last I reached a spot where the bees flew home along a route that was perpendicular to the road, I measured their flight direction: 265°. At this point, I headed into the woods along this same bearing, examining every large tree along or near this line. I did so confident in my knowledge that the bee tree stood somewhere along the line running at 265° from Decker Road to Jackson Creek, a line that was only 150 yards long.

I searched patiently for an hour, but still saw no sign of the bees. It was now after 6:00 P.M., the air was starting to turn cool, and dark thoughts of another failure were starting to fill

FIG. 7.6. Bees entering their hideaway home in the damaged top of a towering hemlock tree.

my mind. My difficulty was that it was impossible to examine the upper regions of these lovely, but exasperatingly tall trees. But then something almost unbelievable happened. While standing along the 265° line, on the crest of the steep slope, I got a call on my mobile phone from Sean Griffin, the Cornell undergraduate student who had helped me with bee hunting in the Arnot Forest back in 2011, when I needed to sample the honey bees living in this forest for studies of their genetics. Sean was calling from Oklahoma and had some questions regarding his master's thesis project on oilseed rape pollination. I was standing there, telling Sean where I was, what I was doing, and how I could sure use his help, when suddenly my eyes, looking west into the treetops that were at eye level, caught the flashing image of sunlit bees zipping in and out of an opening in the top of a huge hemlock tree about 30 yards away (fig. 7.6). I looked more carefully. Yes, those were indeed honey bees! So after three years and 18 days, I had finally found the bee tree. This ancient hemlock tree had suffered an injury to its top, perhaps from a lightning strike or a wind storm, so its topmost section had some deadwood, and in it the bees had found an attractive nesting cavity. It looks like a marvelous place for bees to live: sunny, scenic, and perfectly safe from bears.

It still astounds me when I reflect on the flurry of strokes of fantastic luck that made possible my discovery of the concealed treetop home of these bees. First, the hemlock tree in which they were living was (and still is) growing up from a spot 45 feet below where I stood, so even though the bees' nest entrance proved to be 53 feet up from the base of the tree, it was right at my eye level when I looked out from atop the slope. This was essential to the find, for I could never have detected these bees by looking up from the base of the tree; the countless tree limbs above me would have obscured everything. Second, not only was it necessary that I was standing at the crest of the hill and gazing (unwittingly) west into the nearby treetops, it was also essential that I was standing in *just the right spot*. If I had received Sean's phone call while standing more than 2 yards to the left or to the right of an imaginary line running between me and the bee tree, or more than 2 yards forward or backward along this line, then I would not have spied the bees. Only if you are standing within this 12-foot by 12-foot patch of the forest floor do you have a clear line of sight of the treetop home of these bees. Third, it was a wondrous stroke of good luck that Sean called me when he did, for I later learned, when photographing the nest, that it is only at end of the afternoon, when the sun is in the southwest, that its light shines through a gap in the hemlock boughs, illuminating the bees' flight path and making them shine brightly before their nest. When I reflect on these things, I figure that in finding this bee tree I must have used up whatever was remaining in my lifetime supply of good luck as a bee hunter.

ON FAILURE IN BEE HUNTING

The greatest thrill of a bee hunt comes when, at last, you find the bee-tree home of your bees. Unfortunately, you won't experience this thrill in every hunt. Sometimes you will come home having closely approached, but not found, the bee tree you were seeking. Probably this is good, for isn't it the knowledge that you can easily fail in an endeavor, despite giving it your best efforts, that makes it so sweet when you achieve success?

The Oxford English Dictionary gives four definitions of failure, but it is the second one that relates best to bee hunting: "The fact of becoming exhausted or running short, giving way under trial . . . etc." As a bee hunter, you go out into the countryside, and before you stretches a forest with thousands of trees. Maybe only one is a bee tree. Equipped with your small kit of bee-hunting tools, plus your skill and knowledge as a bee hunter, you establish a beeline; you trail the bees in a series of moves across the fields, up the mountain, and over the swamps; and you whittle down the away times of your bees. Sometimes, your hunt will be cut short by forces beyond your control; you will be shut down by a rainstorm or you will be called away. These are just times of bad luck, not failure. Other times, though, you will be drawn by the bees into deep woods where they circle up from your comb without revealing the way to their home, or you will get close to the bees' home but not find its exact location. Your feelings of optimism and suspense will turn to ones of frustration and failure, for you will have given the hunt your best effort but the challenge was too great. You will, in the words of the OED, experience the fact of "running short." But do take heart, for you have exercised your muscles and mind, probably you have enjoyed places of great natural beauty, and certainly you have engaged the most intelligent insect in the world.

BIOLOGY BOX 7

What Do Honey Bees Seek in a Homesite?

In many animal species, individuals carefully choose a certain microhabitat in which they will build their nests and rear their offspring. This behavior is highly beneficial since the habitat can provide protection from harsh physical conditions and defense from predators. Honey bees conduct an extremely sophisticated process of nest-site selection. No fewer than six distinct properties of a potential homesite—including cavity volume, entrance size, entrance height, and presence of combs from an earlier colony—are assessed to produce an overall judgment of a site's quality. Nest-site selection by honey bees is further intriguing because it is a social process, one that involves several hundred bees simultaneously scouting the environment in a coordinated hunt for the best available dwelling place. This massive search operation often reveals 20 or more potential homesites; only one of these is finally chosen for habitation, and it is almost always the best one (Seeley 2010).

My first step as I sought to understand what honey bees seek in a dwelling place was a descriptive study of the natural nests of honey bees (Seeley and Morse 1976). In this study, we located 21 wild colonies living in hollow trees in the forests around Ithaca, New York. I then felled the bee trees, collected the colonies, and dissected their nests. This work revealed that when honey bees live in the wild, they do so quite differently from when they inhabit a beekeeper's hive. Whereas a beekeeper desires a large, non-swarming colony that is capable of stockpiling a vast quantity of honey (much more honey, in fact, than the colony will ever need), wild colonies have only one-third to one-half the population size, sequester only as much honey as they need, and devote their remaining energies to colony reproduction, by swarming nearly every year and rearing loads of drone (male) bees. These differences between wild colonies and beekeepers' colonies are mainly a result of the beekeeping practice of providing colonies with superabundant nesting space. As a result, the colonies managed by beekeepers store away great quantities of honey and tend not to become over-

Two nest boxes used in a study of the nest-site preferences of honey bees. These boxes offer nesting sites that are identical in all ways (same cavity volume and shape, same entrance height and direction, etc.) except that the one on the right has a smaller entrance opening (2 square inches) than the one on the left (12 square inches).

crowded and swarm. Wild colonies, in contrast, occupy cavities that are usually only a quarter to half the size of a beekeeper's hive, so they easily outgrow this space and swarm each year. Wild colonies also generally live in tree cavities high off the ground, whereas beekeepers' colonies live, of course, in hives set at ground level.

Having found these striking differences between the bee-tree and bee-hive homes of honey bee colonies, I wondered whether they exist because colonies living in the wild choose to live in tree cavities that are relatively small, high up, etc., or if they do so because such cavities are simply what are commonly available in nature. To analyze the nest-site preferences of honey bees, I set out more than 250 nest boxes in pairs around Ithaca and saw which ones were occupied by wild swarms of honey bees (Seeley and Morse 1978). The boxes in each pair were spaced about 30 feet apart on pairs of trees or power-line poles, where they were matched in visibility, wind exposure, and the like (see fig.). Each pair of nest boxes was designed to test one nest-site preference. I did so by giving swarms a choice between one box whose properties all matched those of a typical nest site in nature (typical entrance area, cavity volume, etc.) and one box identical to the first except in one property, the value of which was atypical. Wild swarms could demonstrate a preference in the variable that differed between the boxes in a pair by preferentially occupying one or the other type of box. For example, to test for a preference in entrance height, I set up pairs of nest boxes that were identical except that one was high (16 feet, typical in nature) off the ground and one that was low (3 feet, atypical) to the ground.

As is shown in Table 7.1, the swarms demonstrated preferences in the following nest-site variables: entrance size, entrance direction, entrance height off the ground, entrance height above the cavity floor, cavity volume, and presence of combs in the cavity. The four preferences regarding the entrance opening probably help a colony survive against the threats of cold winters and dangerous predators. A small entrance is easily defended and helps isolate the nest from the outside environment. An entrance high up in a tree is less apt to be discovered by a predator than one near the ground, and it is certainly inaccessible to predators that cannot fly or climb trees. An entrance at the bottom of the nest cavity rather than at the top helps minimize the loss of heat from the colony by convection currents. And an entrance that faces south provides a warm, solar-heated porch from which the foragers can take off and land. The preference of cavities larger than 0.35 cubic feet almost certainly reflects a colony's need for sufficient storage space to hold the honey it needs to survive winter. In cold-climate regions, a honey bee colony needs about 40 pounds of honey-filled combs to fuel its heat production throughout a winter. A honey reserve of this size requires a minimum of about 1.0 cubic foot of nesting space.

TABLE 7.1. Nest-site Properties and Honey Bee Preferences,
Based on Nest-box Occupations by Swarms

A > B, denotes A preferred to B; A = B denotes no preference between A and B.

Property	Preference	Reason
Size of entrance	2 > 12 square inches	Colony defense and thermoregulation
Direction of entrance	South > north facing	Colony thermoregulation
Height of entrance	15 > 3 feet	Colony defense
Position of entrance	Bottom > top of cavity	Colony thermoregulation
Shape of entrance	Circle = vertical crevice	No functional difference
Volume of cavity	0.35 < 1.40 > 3.5 cubic feet	Storage space for honey and colony
Combs in cavity	With > without	Economy in nest construction
Shape of cavity	Cubical = tall	No functional difference
Dryness of cavity	Wet = dry	Bees can waterproof a damp cavity
Draftiness of cavity	Drafty = tight	Bees can caulk cracks and holes

 CHAPTER 8

On Not "Taking Up" the Bee Tree

Traditionally, the final stage in a bee hunt is "taking up" the bee tree—that is, cutting down the tree, splitting open the nest cavity, and taking the bees' honey. This nasty work was often performed directly upon locating the bee tree, but the savvy bee hunter waited until the frost had killed the flowers and the bees had stored up as much honey as they could for the year. Having received permission from the tree's owner, often in exchange for a jar or two of honey, the bee hunter sallied forth with a small crew of woodsmen and perhaps some guests. This company was equipped with a crosscut saw or a chainsaw, an axe, a sledgehammer, at least three stout steel wedges, beekeeper's veils and gloves, a bee smoker, an old kitchen knife or two, and various buckets, wash boilers, and other receptacles for the honeycombs. The bee tree, marked with the hunter's initials, was easily found by following the trail he had blazed with his hand axe while lining the bees. Upon reaching the tree, the bee hunter might daringly climb it and chop open the hollow occupied by the bees. More commonly, though, he and his helpers notched the tree and then sawed a back cut to bring down the entire tree, which fell in a thunderous crash, often splitting open the hollow occupied by the bees. Now the bees poured forth, stinging fiercely until the bee hunter rushed in and puffed

ROBBING A WILD-BEE HIVE.—DRAWN BY R. F. ZOGBAUM.

FIG. 8.1. Illustration titled "Robbing a wild-bee hive," by Rufus Fairchild Zogbaum. It appeared in the November 3, 1883, issue of the magazine *Harper's Weekly* (vol. 27, no. 1402).

smoke in all the apertures, which induced the bees to calm themselves by gorging on the honey oozing from their shattered combs.

With the battle won, the doughty honey hunters next made deep saw cuts across the top of the fallen tree's trunk, above and below the entrance to the bees' home. When honey and wax appeared on the crosscut saw's blade or spit from the chain saw's chain, they knew they had struck the nest. Next, a wedge was held at the base of one of the saw cuts and driven home with the sledgehammer. Then a second was driven in farther down the trunk along its grain, and then perhaps a third along the same line as the first two. Eventually, a slab of the hollow tree's trunk over the nest would break free enough so that it could be pried off like a heavy lid, exposing the bees and their combs of honey. These usually lay in layers, running along for anywhere from 2 to 5 feet inside the tree cavity. Now the honey hunters sliced the combs free from the tree cavity's walls and cut the long slabs of honeycomb into chunks that fit in their buckets. Finally, they would retreat 100 or so yards from the disemboweled bee tree, peel off gloves and veils, and feast on the marvelously delicious fresh honey. Any honeycombs not eaten on site were taken home, broken up, and put into cheesecloth bags hung over pans in a warm room. A day or two later, once the crushed honeycombs had become just dry bits of beeswax, the honey was drained from the pans into glass jars for storage. Thus the bee hunt, turned deadly honey hunt, came to a close.

THE CRUELTY OF TAKING UP A BEE TREE

There is one inescapable fact about the taking up of a bee tree: the cruelty of the act. Once the tree is felled, the nest hollow broached, and the honey removed, the bees will die. Deprived of their food and shelter in the late autumn, the bees have no opportunity to gather more of the former, even if they manage to replace the latter by occupying another tree hollow. The bees have worked hard all summer amassing a stockpile of honey that should be their winter heating fuel, but now they have been pitilessly robbed of the fruits of their labor, and soon their very lives.

Fig. 8.2. Taking up a bee tree. *Left:* Felling the bee tree shown in fig. 1.1. *Right:* The author, in July 1975, examining the combs in a bee-tree nest that he will dissect.

Taking up a bee tree is thrilling, but I cannot encourage the practice. Indeed, I hope fervently that most bee hunters will be too tenderhearted to take up the bee trees they locate. After all, doing so kills two living things worthy of respect: a tree that has survived to old age and a bee colony that has thrived entirely on its own. Furthermore, what bee hunter with a good heart can want to hurt the friendly bees that stayed with him during the hunt, made countless trips to his feeder, and pointed the way forward with their homeward flights? It is certainly far more humane to obtain honey by purchasing it from beekeepers, for they can harvest honey from their colonies in ways that inflict no suffering on the bees.

I raise the ethical issues related to taking up bee trees because I know, from personal experience, that they can be overlooked. Forty summers ago, when I was a young man striving to establish myself as a professional biologist, I took up 21 bee trees (fig. 8.2). I did this in the "name of science"—that is, to achieve the scientifically valid goal of making the first detailed

FIG. 8.3. Completed "cut out" operation. The bee tree on the left had blown over in a gale, exposing at its base the nest of a honey bee colony. All the combs of this colony were cut out and installed in wooden frames that were set inside the bee hive. The bees are gradually moving into the hive, their new home.

description of the natural nests of *Apis mellifera*. This was an important first step in expanding our then scanty knowledge of how honey bees live in the wild, rather than in beekeepers' hives, so I was keen to conduct this study.

Were I to conduct the same investigation today, I would do it with greater concern for the bees. I would collect the data—on cavity size and shape, total comb area, colony population, honey stores, and so forth—by performing a "cut out" at each bee tree. This is what a beekeeper does to extract a colony living in the wall of a building (fig. 8.3). It consists of removing some of the building's siding to fully expose the nest, cutting out the bee-covered combs, securing them with string in wooden frames, installing the comb-filled frames in a bee hive, and finally positioning the hive so its entryway is close by the colony's previous nest entrance. Come nightfall, when all the bees will be inside their new home, the beekeeper returns to screen shut the hive and then move it and the bees to a location where they can live without being a problem. This way, the bees have a fighting chance to put their nest and lives back together, the homeowner has his problem solved and feels virtuous for not killing the bees, and the beekeeper has gained a colony of bees and usually some cash for his service. It is a genuine win-win-win solution to the problem of honey bees living where they are not welcome.

WILD HONEY BEE COLONIES ARE A VALUABLE GENETIC RESOURCE

Why should we value, and even protect, the wild colonies of honey bees with whom we share this planet? To answer this, I find it useful to adopt a historical perspective. To begin, we know from the fossil record that honey bees have been in existence for at least 30 million years, so it is possible that humans have treasured wild colonies of honey bees as a source of delightful food for as long as our species has existed, approximately 150,000 years. It is certainly true that before the tools and techniques of modern beekeeping were developed in the late 1800s, wild colonies were our primary source of honey, the most delicious of all natural foods. These days, though, very little honey comes from the wild. Of the thousands of millions of pounds of honey that mankind sumptuously devours each year, virtually all comes from the 10+ million managed colonies of honey bees that are owned and operated by beekeepers.

Although the honey bees living in the wild are no longer important to us as honey makers, they do remain valuable for their pollination services, worth many tens of billions of dollars worldwide. This may be surprising, but not if you recognize that it is not just the bees flying from beekeepers' hives that pollinate our apple orchards, tomato fields, cranberry bogs, and other croplands. Honey bees from wild colonies—together with bumble bees, solitary bees, and diverse non-bee pollinators—also contribute mightily to the business of agriculture, especially in landscapes that combine fields, woods, and waste areas. These are the habitats that provide plentiful nesting sites and food sources for wild honey bees and all the other free-service pollinators.

There is one more way, besides pollination and (very limited) honey production, in which wild colonies of honey bees are of value to humanity these days: as a special genetic resource for the species *Apis mellifera* as it adapts to threats of disease. Over the past 50 or so years, the honey bee has joined the list of species threatened by the emergence of new infectious diseases. The main cause of this peril for honey bees is the global spread of a species of mite named *Varroa destructor*. These pinhead-size, frisbee-shaped mites are native to Asia, where they parasitize, but rarely kill, colonies of the eastern hive honey bee, *Apis cerana,* which lives only in Asia. Unfortunately, humans accidentally introduced these mites to the western hive honey bee, *Apis mellifera,* which originally lived in just Africa, the Middle East, and Europe but now lives in all the continents except Antarctica. This host shift by the mites matters greatly because *Varroa destructor*, as its name implies, is no friend of its new host species. Worldwide, this mite has contributed to the deaths of millions of *Apis mellifera* colonies. *Varroa* is so deadly because it is so fiendishly effective at transmitting from bee to bee the many viruses to which bees are vulnerable, some species of which have extremely lethal strains. The deformed-wing virus (DWV), for example, is highly injurious because worker bees heavily infected with it develop shriveled wings and so cannot fly. Shortly after a colony's population of the *Varroa* mites explodes, its population of worker bees implodes, as they die from infections of DWV and other viruses. Beekeepers have dubbed this phenomenon Colony Collapse Disorder (CCD).

The main way that beekeepers in North America and Europe have responded to the heavy die-offs of their colonies—30% or more per year—caused by viruses vectored by *Varroa* is to regularly treat their colonies with pesticides that kill the mites. This approach, however, is not a sustainable remedy for CCD for four reasons: it promotes the evolution of resistance by the mites to these chemicals, it can contaminate the honey crop with pesticides, and it can have negative effects on the bees themselves. But probably the greatest shortcoming to repeatedly dosing colonies with pesticides is that it blunts the process of natural selection for bees with resistance to the mites and the viruses.

Honey bee colonies living on their own in the wild receive no treatments with pesticides for killing mites, so they must rely instead on their inherent abilities to resist diseases. This means that populations of wild colonies experience relentless selection for resistance to the deadly *Varroa*-DWV association. And not just in theory. We now know that natural selection has given rise to mite resistance in at least three populations of wild (unmanaged) colonies of *Apis mellifera*: in Russia, Sweden, and France. The Russian bees are skilled at thwarting the mite's reproduction by dragging from the nest most of the developing bees (pupae) that are parasitized by a female mite. This lowers the mites' reproductive success and so suppresses the mite population within a colony. In the Swedish and French populations of mite-resistant bees, mite reproduction is also inhibited, but how the Swedish and French bees are achieving this lifesaving measure remains a mystery.

What about the population of wild colonies living in and around the Arnot Forest? I don't know (yet) if this population of wild honey bee colonies has evolved genetic traits that are helping them survive on their own. What I do know, though, is that the colonies living in this forest are infected with *Varroa* mites, are not receiving pesticide treatments to control their mite infestations, and yet are surviving and reproducing well enough to sustain this population of colonies. I also know from genetic analyses that this population of wild colonies is *not* bolstered by immigration of swarms from the (very few) bee-keepers' colonies in the area. What is most intriguing about the Arnot Forest bees, however, is what has been learned from a recent study that compared the genomes of worker honey bees collected from colonies in this forest in 1977 and 2011: this population of colonies suffered an episode of massive mortality sometime in the 34-year period between 1977 and 2011. This die-off probably occurred shortly after the arrival of *Varroa destructor* in New York State, hence in the mid-1990s. The unmistakable signature of high colony death in the Arnot Forest , which created a "population bottleneck" there, is a stunning loss of diversity in these bees' genes between 1977 and 2011. All the colonies living in this forest today are descendants of just a handful of colonies, maybe only three or four, that survived the arrival of *Varroa destructor*. This study has also revealed that hundreds of genes in the Arnot Forest bees show signs of strong selection, which is to be expected for a species that has suffered high mortality from a new pathogen or parasite. The survivors in such populations are likely to carry genes that confer resistance to the new disease agents.

More work is needed to understand fully the evolution of disease resistance in populations of wild honey bee colonies, including the one in the Arnot Forest. It now seems clear, however, that if a population of wild colonies is allowed to live on its own (so the bees experience strong natural selection), and if this population is not flooded with genes from elsewhere (so natural selection can increase the frequencies of the beneficial versions of the population's genes), then this population of wild colonies will evolve a balanced relationship with its agents of disease and indeed with its environment as a whole. I suggest, therefore, that as we ponder how to promote the welfare of *Apis mellifera*, our greatest friend among the insects, we should heed the words of the famous naturalist and avid bee hunter, Henry David Thoreau: "In Wildness is the preservation of the world."

BIOLOGY BOX 8
How Can You Acquire Wild Colonies of Honey Bees?

If you have access to a large wooded area that is several miles from any beekeepers' hives, then it is easy to acquire a honey bee colony with the genetics of the wild bees living in this location. You simply put up nest boxes ("bait hives") of the *right design* in the *right location* and at the *right time of year*. I have been capturing swarms in bait hives set in the countryside around Ithaca, New York, for over 40 years, and it is now my primary means of getting new colonies. On average, I capture one swarm each summer for every two or three bait hives that I put out.

The first step toward success with bait hives is to build wooden boxes that will be attractive to honey bees looking for a home. Biology Box 7 summarizes what makes a dream home for honey bees: a cavity with a volume of 40 liters (approx. 1.5 cubic feet) and an entrance opening whose area is about 12 square centimeters (approx. 2 square inches), faces south, is near the bottom of the cavity, and is high off the ground. To create a bait hive that meets these criteria, I take an old beehive and use a block of wood to reduce the entrance opening to about 2 square inches. Next, I fill the hive with frames of old, aromatic beeswax comb. Finally, I mount the hive in a place where it will be at least 10 feet off the ground. A tree house (see fig.) or a porch roof are perfect spots. If possible, I face the entrance south. It is my impression, though I don't have hard data on this, that I have greater success if the bait hive is mounted in a large tree on the edge of the woods, perhaps because this is where a nest-site scout is most likely to seek (in a large tree) and to find (in a conspicuous location) a prospective homesite.

Bait hive installed in a tree house. This is a standard, movable-frame hive that has had its entrance reduced to an opening of approximately 2 square inches. It contains 10 wooden frames filled with old combs. There are no honey bee colonies managed by beekeepers within 2 miles of this tree house, and yet nearly every year a swarm moves into the bait hive installed here.

To be successful with bait hives, it helps to know when swarming takes place in your area. Swarming is the honey bee's process of colony reproduction in which the majority of an established colony's members—a crowd of some 10,000 worker bees—flies off with the mother queen to produce a daughter colony, while the rest stay at home and rear a new queen to perpetuate the parental colony. It happens mainly in the late spring and early summer. This is May–July where I live. Naturally, you will want to have your bait hives set out a few weeks in advance of the swarming season. Once you have them deployed, it is wise to check them every week or so, for you will want to collect the occupied ones before the bees have loaded them up with honey. To take down an occupied bait hive, I first puff some smoke in the entrance; this calms the bees and coaxes them to move inside. Next, I staple a piece of wire screen over the entrance. Finally, I use a rope to lower the hive to the ground. *Note:* do not climb down the ladder while holding a bait hive; this is a very dangerous thing to attempt.

If you build your bait hive properly and position it carefully, and if nature endows your location with plentiful swarms, then you will have a high likelihood of catching a swarm. If this happens, you have my congratulations, for now you are entitled to call yourself not just a bee hunter, but also a bee trapper.

 *Notes*

PREFACE

Page x: The best general reference on humanity's use of honey bees from prehistoric times to the present day is Eva Crane's *The World History of Beekeeping and Honey Hunting* (Crane 1999).

Page x: The earliest known description of bee hunting was written by Lucius Junius Moderatus Columella. It appeared in book 9 (on beekeeping) of his work *De Res Rustica (On Agriculture)*, the most comprehensive and detailed of Roman agricultural works. For a translation, see Columella (1968).

Page x: Two books are excellent sources of information about the long history of bee hunting to find wild colonies of honey bees. Eva Crane's *The World History of Beekeeping and Honey Hunting* (Crane 1999) examines honey and bee hunting worldwide. Dorothy Galton's *Survey of a Thousand Years of Beekeeping in Russia* (Galton 1971) gives a detailed account of the intense beekeeping activity in the Russian forests in the 1100s to the 1600s. It describes the work of the beeman (*bortnik*), who kept his colonies of bees in holes in cavities in the trees of the forest, and who was skilled in finding these wild colonies by bee hunting.

Page x: For drawings showing bee traps—made of wood, a deer's antler, or horn—and the methods of bee hunters in the Carpathian Mountains (which run from the Czech Republic through Slovakia, Poland, Hungary, Ukraine, and Romania); see Gunda (1968).

Page x: Hollow-tree beekeeping is still conducted in the South Ural area of the Republic of Bashkortostan. For superb photos depicting the equipment and methods of this early form of beekeeping, see Ilyasov et al. (2015).

Page xi: Excellent accounts of the origins of beekeeping using fixed-comb hives (in tree sections, skeps, or clay pipes) and its subsequent transition to movable-comb hives (also known as top-bar hives) and movable-frame hives (for example, Langstroth hives) are found in parts VI to VIII of Eva Crane's *The World History of Beekeeping and Honey Hunting* (Crane 1999).

Page xv. There are many explanations of the waggle dance of honey bees. Of these, I recommend the little book written by Karl von Frisch, the Austrian zoologist who decoded the bees' "dance language" (von Frisch 1971).

CHAPTER 1. INTRODUCTION

Page 3: Aldo Leopold quote, from Leopold (1949), p. vii.

Page 5: For more information on the phenomenal sex life of honey bees, see Koeniger et al. (2014).

Page 5: Crane (1983) reviews the abundant archaeological evidence (texts and wall paintings) demonstrating that the honey bee has a long relationship with humans, dating back to at least 7000 B.C., when Egyptian and Mesopotamian beekeepers used bees to produce precious wax and honey. Bloch et al. (2010) describe the discovery of the most ancient bee hives (clay cylinders) yet found in Tel Rehov, in northern Israel. They are approximately 3,000 years old.

Page 5: The remarkable lack of domestication of honey bees is reviewed by Oldroyd (2012) in the ironically titled paper "Domestication of honey bees was associated with expansion of genetic diversity."

Page 5: To read Thoreau's journal entry of September 30, 1852, in its entirety, see Thoreau (1962).

Page 11: The research by myself and many others on how a swarm of honey bees chooses its nesting cavity, through a process of collective intelligence, is reviewed in Seeley (2010).

Page 11: The problems that arise when honey bees live in hives packed together in apiaries are reviewed in Seeley and Smith (2015).

Page 12: For a description and ecological history of the Arnot Forest, see Odell et al. (1980).

CHAPTER 2. THE BEE BOX AND OTHER TOOLS

Page 36: Fine (size 000 or 0000) artist's brushes with red sable or camel's hair bristles work especially well for labeling bees, because they make it easy to apply gently a dot of paint on a bee's thorax or abdomen. The shellac-based paint sets that I like to use with such brushes are described on pp. 79–80 in Seeley (1995). Watercolor paints applied with a small brush work well too. Regarding paint pens, some have foul-smelling solvents that disturb the bees and make labeling them difficult, so these should be avoided. I've had good success labeling bees using water-based paint markers from POSCA; see www.POSCA.com (accessed June 30, 2015).

CHAPTER 3. BEE-HUNTING SEASON

Page 48: For more information on the design and use of bait hives (also known as swarm traps) to capture swarms of honey bees, see Seeley (2012a or 2012b) or Repasky (2013).

Page 49: That honey bees experience a high risk of death when working as foragers outside the hive has been revealed by Visscher and Dukas (1997). They recorded the lifetime foraging activity and survivorship of individual honey bees and discovered that a worker bee has a probability of 0.04 of dying during each hour she spends outside the hive, regardless of her age. This constant probability of mortality for each hour spent foraging indicates that these bees died mainly from accidents and predation, not from old age (senescence).

Page 49: Bartholomew (1981) explains that flapping flight is the most energetically expensive mode of animal locomotion, and he compares the metabolic rates of various animals, including those with flapping flight: birds (for example, hummingbirds) and insects (bees and moths). To see that the flight muscles of insects are the most metabolically active of tissues, consider the weight-specific rates of oxygen consumption for a 2-gram hummingbird (a tiny bird famous for its phenomenal metabolic rate) and a 60-milligram honey bee: 50 cm^3 of oxygen per gram per hour vs. 100 cm^3 of oxygen per gram per hour. Per unit of body weight, a flying worker bee consumes oxygen, to burn calories, twice as fast as a flying hummingbird.

Page 50: For a graph that shows the day-to-day fluctuations in a colony's nectar collection for the period May through July in northeast Connecticut, see fig. 2.15 in Seeley (1995). The episodic honey flows from dandelions, black locust trees, sumac shrubs, and basswood trees are striking.

CHAPTER 4. ESTABLISHING A BEELINE

Page 63: For more information on evaporative cooling of the nest by honey bees, see Kühnholz and Seeley (1997).

Page 67: For more information on how a bee learns the look of your bee box and the arrangement of the landmarks around it when she leaves your box for the first time, see Lehrer and Bianco (2000).

Page 68: The danger inherent in being a robber bee is described vividly in Butler and Free (1952). Robber bees that are detected by a colony's guards are seized, stung, and then die 2–3 minutes later.

Page 74: The brain of a worker honey bee is small, but it contains approximately 1 million neurons, so it is not surprising that these insects show impressive behavioral complexity. For reviews of their cognitive abilities from a neuroscience perspective, see Menzel (2012); from a behavioral perspective, see Seeley (2003b).

CHAPTER 5. TIMING BEES TO ESTIMATE DISTANCE TO HOME

Page 82: For more information on the coordination of the worker bees that collect nectar and those that process nectar within a honey bee colony, see chapter 6 in Seeley (1995). For more information on how honey is produced from nectar, see Crane (1980).

Page 83: A beautiful explanation of how the waggle dance works is found in the little book by the man who deciphered it, von Frisch (1971).

Page 86: The study in which these flight-speed measurements were originally reported is Seeley (1986).

CHAPTER 6. MAKING MOVES DOWN THE BEELINE

Page 106: The reason it is important to stay as close as possible to your bees' flight line home when making moves of your feeding station is that if you move widely off this line, then the bees may become lost. This can happen because worker honey bees do not have a detailed map of the landmarks of the region over which they travel. Instead, they have skill in getting home by dead reckoning. In other words, a foraging bee knows what direction and what distance she is from the hive, and she finds her way home by flying back in the right direction and for the right distance. So if you greatly displace a bee from her normal flight route home, then she is apt to never reach home because the flight direction and flight distance she knows will not get her back to her home base.

Page 111: Over the years, in and around Ithaca, New York, I have observed eight colonies that built their nests out in the open rather than inside a protective cavity. None survived the winter.

CHAPTER 7. FINDING THE BEE TREE

Page 119: Propolis (from Greek: *pro*, before, and *polis*, city; referring to its use around the entrances to honey bee nests) is resin that the bees have gathered from plants, especially from the buds of poplar, birch, horse chestnut, pine, and spruce trees. Bees use propolis to reduce their nest's entrance opening to render it more weather-tight and easier to defend, to create a smooth landing area outside the entrance, and to create a tough, clean coating on the nest cavity's walls. It has strong antimicrobial activity. See Simone-Finstrom and Spivak (2010).

Page 124: The amazing process of group decision making by which the bees in a honey bee swarm choose their future dwelling place is described in Seeley (2010).

Page 131: The genetic studies referred to here are Seeley et al. (2015) and Mikheyev et al. (2015).

CHAPTER 8. ON NOT "TAKING UP" THE BEE TREE

Page 140: You might wonder how much honey one gets from a bee tree. The amount varies enormously. From the bee trees I took up in July and August 1975, I obtained from 0 to 57 pounds, with 29 pounds the average amount. Edgell (1949) reported similar amounts: 0–97 pounds, and an average of about 20 pounds.

Page 141: The published report of this study is Seeley and Morse (1976).

Page 142: A video showing how a "cut out" is done is available at https://www.youtube.com/watch?v=WjLOaYuNcB0 (accessed July 15, 2015).

Page 143: The rich and beautiful fossil record of honey bees (*Apis* spp.) is reviewed broadly in Grimaldi and Engel (2005; see esp. fig. 11.82) and in greater detail in Zeuner and Manning (1976) and Culliney (1983).

Page 143: The statement regarding worldwide honey production is based on figures reported on the websites of the World Trade Daily (www.worldtradedaily.com) and the National Honey Board (www.honey.com/newsroom/press-kits/honey-industry-facts; both accessed July 15, 2015). The statement regarding total number of managed honey bee colonies worldwide is based on fig. 1 in van Engelsdorp and Meixner (2010).

Page 144: For 2009, in the United States alone, the commercial values attributed to crop pollination by honey bees and non-honey-bee pollinators were estimated to be $11.7 billion and $3.4 billion, respectively. See Calderone (2012).

Page 144: Rosenkranz et al. (2010) review how *Varroa destructor* moved from its original host *Apis cerana*, an Asian honey bee, to *Apis mellifera* when colonies of the latter were imported to Asia, and how these mites, by spreading viruses, are the most detrimental honey bee parasite and the largest threat to beekeeping. See also Martin et al. (2012) for how this mite has greatly increased the prevalence of one especially harmful strain of the deformed wing virus in honey bee colonies.

Page 144: *Varroa* mites function in two ways as an effective vector of the bees' viruses: (1) by acting as a reservoir, and perhaps even an incubator, of the viruses, and (2) by then spreading these viruses by injecting them into the bees' blood when the mites feed on the bees.

Page 145: Locke (2016) reviews what is known about populations of *Apis mellifera* that have evolved resistance to *Varroa destructor*.

Page 146: See Seeley et al. (2015) and Mikheyev et al. (2015) for the genetic studies of the wild colonies of honey bees living in the Arnot Forest.

References

Bartholomew, G. A. 1981. "A matter of size: an examination of endothermy in insects and terrestrial vertebrates." In *Insect Thermoregulation*, B. Heinrich, ed., Wiley, New York, pp. 45–78.

Bloch, G., T. M. Francoy, I. Wachtel, N. Panitz-Cohen, S. Fuchs, and A. Mazar. 2010. "Industrial apiculture in the Jordan valley during Biblical times with Anatolian honeybees." *Proceedings of the National Academy of Sciences of the United States of America* 107: 11240–11244.

Brant, C. 1966. *The Bee Hunter*. Pageant, New York.

Burroughs, J. S. 1875. *Birds and Bees*. Houghton Mifflin, New York.

Butler, C. G., and J. B. Free. 1952. "The behaviour of worker honeybees at the hive entrance." *Behaviour* 4: 262–292.

Calderone, N. W. 2012. "Insect pollinated crops, insect pollinators, and US agriculture: trend analysis of aggregate data for the period 1992–2009." PLoS ONE 7:e37235.

Cartwright, B. A., and T. S. Collett. 1982. "How honey bees use landmarks to guide their return to a food source." *Nature* 295: 560–564.

Columella, L. J. M. 1968. *On Agriculture: Books V–IX*. Translated from the Latin by E. S. Forster and E. H. Heffner. Harvard University Press, Cambridge, Massachusetts.

Cooper, J. F. 1848. *The Oak Openings; or, The Bee-hunter*. Burgess, Stringer and Co., New York.

Crane, E. 1980. *A Book of Honey*. Scribner, New York.

———. 1983. *The Archaeology of Beekeeping*. Duckworth, London.

———. 1999. *The World History of Beekeeping and Honey Hunting*. Routledge, New York.

Culliney, T. W. 1983. "Origin and evolutionary history of the honeybees *Apis*." *Bee World* 64: 29–38.

Donovan, R. E. 1980. *Hunting Wild Bees*. Winchester Press, Tulsa, Oklahoma.

Dudley, P. 1720. "An account of a method lately found out in New-England, for discovering where the bees hive in the woods, in order to get their honey." *Philosophical Transactions of the Royal Society of London* 31: 148–150.

Edgell, G. H. 1949. *The Bee Hunter*. Harvard University Press, Cambridge, Massachusetts.

Galton, D. 1971. *Survey of a Thousand Years of Beekeeping in Russia*. Bee Research Association, London.

Grimaldi, D., and M. S. Engel. 2005. *Evolution of the Insects*. Cambridge University Press, New York.

Gunda, B. 1968. "Bee-hunting in the Carpathian area." *Acta Etnográfica Hungarica* 17: 1–62.

Hinson, E. M., M. Duncan, J. Lim, J. Arundel, and B. P. Oldroyd. 2015. "The density of feral honey bee (*Apis mellifera*) colonies in South East Australia is greater in undisturbed than in disturbed habitats." *Apidologie* 46: 403–413.

Ilyasov, R. A., M. N. Kosarev, A. Neal, and F. G. Yumaguzhin. 2015. "Burzyan wild-hive honeybee *A. m. mellifera* in South Ural." *The Beekeepers Quarterly* 119: 25–33.

Irving, W. 1835. *A Tour on the Prairies*. Carey, Lea, and Blanchard, Philadelphia.

Koeniger, G., N. Koeniger, J. Ellis, and L. Connor. 2014. *Mating Biology of Honey Bees (Apis mellifera)*. Wicwas Press, Kalamazoo, Michigan.

Kühnholz, S., and T. D. Seeley. 1997. "The control of water collection in honey bee colonies." *Behavioral Ecology and Sociobiology* 41: 407–422.

Lehrer, M., and G. Bianco. 2000. "The turn-back-and-look behaviour: bee vs. robot." *Biological Cybernetics* 83: 211–229.

Leopold, A. 1949. *A Sand County Almanac, and Sketches Here and There*. Oxford University Press, New York.

Lockard, J. R. 1908. *Bee Hunting*. Harding, Columbus, Ohio. Available as a free download from Project Gutenberg, http://www.gutenberg.org/ebooks/34044?msg=welcome_stranger.

Locke, B. 2016. "*Varroa* mite-resistant *Apis mellifera* honey bees." *Apidologie*. In press.

Martin, S. J., A. C. Highfield, L. Brettell, E. M. Villalobos, G. E. Budge, M. Powell, S. Nikaido, and D. C. Schroeder. 2012. "Global honey bee viral landscape altered by a parasitic mite." *Science* 336: 1304–1306.

Menzel, R. 2012. "The honeybee as a model for understanding the basis of cognition." *Nature Reviews Neuroscience* 13: 758–768.

Mikheyev, A. S., M.M.Y. Tin, J. Arora, and T. D. Seeley. 2015. "Museum samples reveal rapid evolution by wild honey bees exposed to a novel parasite." *Nature Communications*. doi: 10.1038/ncomms8991

Morse, J. 1931. *Following the Bee Line*. Thomas S. Rockwell, Chicago.

Moulton, G. E. 2002. *The Definitive Journals of Lewis and Clark.* University of Nebraska Press, Lincoln, Nebraska.

Odell, A. L., J. P. Lassoie, and R. W. Morrow. 1980. "A history of Cornell University's Arnot Forest. Dept. of Natural Resources Research and Extension," Ser. 14, 1–53. http://www2.dnr.cornell.edu/arnot/about/history.htm (accessed July 15, 2015).

Oldroyd, B. P. 2012. "Domestication of honey bees was associated with expansion of genetic diversity." *Molecular Evolution* 21: 4409–4411.

Repasky, S. J. 2013. *Swarm Essentials.* Wicwas Press, Kalamazoo, Michigan.

Rosenkranz, P., P. Aumeier, and B. Ziegelmann. 2010. "Biology and control of Varroa destructor." *Journal of Invertebrate Pathology* 103: S96–S119.

Seeley, T. D. 1986. "Social foraging by honeybees: how colonies allocate foragers among patches of flowers." *Behavioral Ecology and Sociobiology* 19: 343–354.

———. 1994. "Honey bee foragers as sensory units of their colonies." *Behavioral Ecology and Sociobiology* 34: 51–62.

———. 1995. *The Wisdom of the Hive.* Harvard University Press, Cambridge, Massachusetts.

———. 2003a. "Bees in the forest, still." *Bee Culture* 131 (January): 24–27.

———. 2003b. "What studies of communication have revealed about the minds of worker honey bees." In *Genes, Behaviors, and Evolution of Social Insects,* T. Kikuchi, N. Azuma, and S. Higashi, eds., 21–33. Hokkaido University Press, Sapporo.

———. 2007. "Honey bees of the Arnot Forest: a population of feral colonies persisting with *Varroa destructor* in the northeastern United States." *Apidologie* 38: 19–29.

———. 2010. *Honeybee Democracy.* Princeton University Press, Princeton, New Jersey.

———. 2012a. "Using bait hives." *Bee Culture* (April): 73–75.

———. 2012b. "Capturing swarms with bait hives." *The Beekeepers Quarterly* (March): 33–35.

Seeley, T. D., and R. A. Morse. 1976. "The nest of the honey bee (*Apis mellifera* L.)." *Insectes Sociaux* 23: 495–512.

———. 1978. "Nest site selection by the honey bee, *Apis mellifera.*" *Insectes Sociaux* 25: 323–337.

Seeley, T. D., and M. L. Smith. 2015. "Crowding honeybee colonies in apiaries can increase their vulnerability to the deadly ectoparasitic mite *Varroa destructor.*" *Apidologie* 46: 716–727.

Seeley, T. D., D. R. Tarpy, S. R. Griffin, A. Carcione, and D. A. Delaney. 2015. "A survivor population of wild colonies of European honeybees in the northeastern United States: investigating its genetic structure." *Apidologie* 46:654-666.

Sheppard, W. S. 1989. "A history of the introduction of honey bee races into the United States." *American Bee Journal* 129: 617–619, 664–666.

Simone-Finstrom, M., and M. Spivak. 2010. "Propolis and bee health: the natural history and significance of resin use by honey bees." *Apidologie* 41: 295–311.

Smith, A. J. 2010. *The Appalachian Chronicles. Rural American Philosophy and Folklore from Beelining to Back Porches.* PublishAmerica, Baltimore.

Thompson, W. W. 1925. *Historical sketches of Potter County, Pennsylvania.* The Potter Enterprise, Coudersport, Pennsylvania.

Thoreau, H. D. 1962. *The Journal of Henry D. Thoreau.* B. Torrey and F. H. Allen, eds. Dover Publications, New York.

vanEngelsdorp, D., and M. D. Meixner. 2010. "A historical review of managed honey bee populations in Europe and the United States and the factors that may affect them." *Journal of Invertebrate Pathology* 103: S80–S95.

Visscher, P. K., and R. Dukas. 1997. "Survivorship of foraging honey bees." *Insectes Sociaux* 44: 1–5.

Visscher, P. K., and T. D. Seeley. 1982. "Foraging strategy of honeybee colonies in a temperate deciduous forest." *Ecology* 63: 1790–1801.

von Frisch, K. 1967. *The Dance Language and Orientation of Bees.* Harvard University Press, Cambridge, Massachusetts.

———. 1971. *Bees: Their Vision, Chemical Senses, and Language.* Cornell University Press, Ithaca, New York.

Wehner, R., and R. Menzel. 1990. "Do insects have cognitive maps?" *Annual Review of Neuroscience* 13: 403–414.

Wray, M. K., B. A. Klein, H. R. Mattila, and T. D. Seeley. 2008. "Honeybees do not reject dances for 'implausible' locations: reconsidering the evidence for cognitive maps in insects." *Animal Behaviour* 76: 261–269.

Zeuner, F. E., and F. J. Manning. 1976. "A monograph on fossil bees (Hymenoptera: Apoidea)." *Bulletin of the British Museum of Natural History (Geology)* 27: 149–268.

 *Illustration Credits*

Fig. 1.1. Photo by Thomas D. Seeley
Fig. 1.2. Photo by Thomas D. Seeley
Fig. 1.3. Photo provided by the Pierpont Morgan Library, New York. MA 1302.19. Purchased by Pierpoint Morgan with the Wakeman Collection, 1909.
Fig. 1.4. Photo by Alexander L. Wild
Fig. 1.5. Title page of *The Bee Hunter* by George Harold Edgell, Cambridge, Mass.: Harvard University Press, Copyright © 1949 by George Harold Edgell
Fig. 1.6. Photo by Thomas D. Seeley
Fig. 1.7. Photo by Thomas D. Seeley
Fig. 1.8. Photo by Thomas D. Seeley
Fig. 1.9. Photo by Thomas D. Seeley
Biology box 1 fig.: Original drawing by Margaret C. Nelson
Fig. 2.1. Photo by Megan E. Denver
Fig. 2.2. Original drawing by Margaret C. Nelson
Fig. 2.3. Photo by Megan E. Denver
Fig. 2.4. Photo by Megan E. Denver
Fig. 2.5. Photos by Megan E. Denver
Fig. 2.6. Photo by Megan E. Denver
Biology box 2 fig.: Original drawing by Margaret C. Nelson
Fig. 3.1. Photo by Jorik Phillips
Fig. 3.2. Photo by Helga R. Heilmann
Fig. 3.3. Photo by Megan E. Denver
Biology box 3 fig. A: Original drawing by Margaret C. Nelson
Biology box 3 fig. B: Original drawing by Margaret C. Nelson
Fig. 4.1. Original drawing by Margaret C. Nelson
Fig. 4.2. Photo by Helga R. Heilmann
Fig. 4.3. Photos by Megan E. Denver
Fig. 4.4. Photo by Helga R. Heilmann

Fig. 4.5. Photos by Megan E. Denver
Fig. 4.6. Photo by Helga R. Heilmann
Fig. 4.7. Photo by Megan E. Denver
Fig. 4.8. Photo by Thomas D. Seeley
Biology box 4 fig.: Original drawing by Margaret C. Nelson
Fig. 5.1. Photo by Megan E. Denver
Fig. 5.2. Photo by Kenneth Lorenzen
Fig. 5.3. Photo by Helga R. Heilmann
Fig. 5.4. Photo by Thomas D. Seeley
Fig. 5.5. Original drawing by Margaret C. Nelson
Fig. 5.6. Photo by Thomas D. Seeley
Fig. 5.7. Photo by Thomas D. Seeley
Biology box 5 fig.: Original drawing by Margaret C. Nelson
Fig. 6.1. Photo by Thomas D. Seeley
Fig. 6.2. Original drawing by Margaret C. Nelson
Fig. 6.3. Photo by Thomas D. Seeley
Fig. 6.4. Photo by Thomas D. Seeley
Fig. 6.5. Modified from the figure in P. Dudley, 1720, "An account of a method lately found in New-England, for discovering where the bees hive in the woods, in order to get their honey," *Philosophical Transactions of the Royal Society of London* 31: 148–150.
Biology box 6 fig.: Modified from fig. 16 in B. A. Cartwright and T. S. Collett, 1982, "Landmark learning in bees," *Journal of Comparative Physiology* A 151: 521–543.
Fig. 7.1. Left, photo by Thomas D. Seeley; right, photo by Megan E. Denver
Fig. 7.2. Photo by Thomas D. Seeley
Fig. 7.3. Photo by Thomas D. Seeley
Fig. 7.4. Photo by Thomas D. Seeley
Fig. 7.5. Photo by Megan E. Denver
Fig. 7.6. Photo by Thomas D. Seeley
Biology box 7 fig.: Photo by Thomas D. Seeley
Fig. 8.1. Photo by Megan E. Denver
Fig. 8.2. Photos by Thomas D. Seeley
Fig. 8.3. Photo by Megan E. Denver
Biology box 8 fig.: Photo by Thomas D. Seeley

 *Index*